Content

"You provide me with another mind
as well as another pair of hands."

A class teacher to her teaching assistant

BE A Mathematician
BEAM

Assisting Numeracy

A handbook for teaching assistants
– 2nd edition –

Ruth Aplin and Fran Mosley

BEAM would like to thank the following for their help in the development of this book:

Rob Briscoe   Anne Kennelly
Sheila Ebbutt   Barbara Newmarch
Ros Elphinstone   Billie Old
Sarah Gale   Anita Straker
Ian Grant

Hasan Chawdhry and Wilberforce Primary School, Westminster
Sandra Roberts and Shapla Primary School, Tower Hamlets
Henrietta Harrison and Smithy Street Primary School, Tower Hamlets
Jackie Trudgeon and Canon Barnett Primary School, Tower Hamlets
Sue Walsh and Ben Jonson Primary School, Tower Hamlets
The learning support assistants in Tower Hamlets who commented on and trialled *Assisting Numeracy*
The BEAM Development Group

Published by BEAM Education
Maze Workshops
72a Southgate Road
London N1 3JT

Telephone 020 7684 3323
Fax 020 7684 3334
Email info@beam.co.uk

www.beam.co.uk

ISBN 978 1 9070 3400 8

British Library Cataloguing-in-Publication Data
Data available

Edited by Marion Dill
Cover, design and layout by Matt Carr
Illustrations by Matt Carr
Photos © grantlylynch.co.uk

Printed by Graphycems, Spain

This book was a joint venture between BEAM, the National Numeracy Project and the London Borough of Tower Hamlets.

1 Introduction

Assisting Numeracy is for learning support assistants – that is, teaching assistants – and the teachers they work with. The book looks at the ways in which mathematics is taught in schools today and provides practical support in the form of maths games and activities.

The introduction that follows is divided into two parts. The first part is for the teaching assistant. The second part is for the teacher who works with the teaching assistant or is involved in their training.

To the teaching assistant

Numeracy

We all want the children we work with to grow up and thrive in the modern world. In order to thrive, children need to be able to use numbers and mathematics more generally in all sorts of ways and in every kind of situation. Teachers and teaching assistants both have important parts to play in helping children learn how to use numbers flexibly and accurately.

Numerate childen have more than just a good knowledge of numbers. They are able to apply this knowledge of numbers in real-life situations.

Numerate childen:

- have a feel for the size of a number
- know where numbers fit in the number system
- can rapidly recall the times tables and other number facts
- use what they rapidly recall to work things out in their head
- use a range of methods of calculating in their head and on paper
- use a calculator when they really need to
- make sense of number problems and know how to solve them
- check their answers
- know whether their answer is reasonable
- talk about how they work things out
- explain their mathematical ideas.

What you can do

All children can become numerate, but they need help to get there. There are various ways in which you can help children get better at working with numbers.

You can:

- interest and motivate children
- help them understand how to do an activity
- hear them rehearse number facts
- give them confidence to work something out on their own
- listen to their difficulties when they are doing something challenging
- ask them questions to get them thinking
- help them work together sociably
- encourage them to play
- support them in doing practical activities

and above all

- give them the chance to discuss the mathematics that they are doing.

The job you do

Although the role of the teaching assistant varies from school to school, there are many jobs that all assistants have in common. You will probably recognise many of the jobs listed below as things you have been asked to do in the past, or still do now.

Job description for a teaching assistant

Job purpose
To assist the class teacher with the education, supervision and welfare of all the children in the class

Duties
The particular responsibilities of the post holder will be to:

- work with small groups of children under the teacher's supervision. For example:
  - assisting children when carrying out practical science or mathematics tasks, using equipment
  - assisting children with meanings of words, spelling, handwriting, presentation
  - carrying out practice and revision activities with children
  - supervising the playing of educational games
  - hearing children read
- work with children with varying degrees of learning difficulties
- work with children for whom English is an additional language
- assist in the long- and short-term planning of work with children
- discuss children's progress with teachers
- prepare work and activities for children in advance, in accordance with the objectives set by the teacher
- help children in individual and collaborative study skills
- give children feedback on their achievements and progress

- organise and supervise children in the playground
- operate electronic equipment
- make books, labels and signs and undertake any other practical tasks to maintain a good standard in classroom appearance and organisation
- meet children's physical needs while encouraging independence. For example:
  - help children change for PE lessons or swimming
  - clean and reassure children after accidental soiling of clothes
- administer First Aid as needed and accompany sick or injured children home
- help ensure that a high standard of health and safety is maintained at all times
- help ensure that children adhere to the school's behaviour policy
- accompany teachers and children on educational visits
- attend INSET days and staff meetings
- be familiar with school policies and schemes of work.

If there are some things on this list for which you are not currently responsible, but which you would like to try out and take on, show the list to your headteacher, or whoever is your manager, and ask whether your role might be renegotiated to include these extra responsibilities.

Supporting children with particular needs

Teaching assistants often help children who have individual educational needs. These may be children with physical disabilities, children who have difficulties learning at the same rate as others of their age or children who are learning English as an additional language.

The teacher will plan the mathematics curriculum that these children follow, but children with individual needs often benefit from extra adult attention and intervention. The teacher needs to liaise closely with you when you are working for long periods with specific children, to make sure that the children are progressing in the way that you and the teacher intend. There is a delicate balance between giving crucial support to children and finding that they have become dependent on adult attention. Discussing individual children's learning with the teacher is an important part of your role, and the school should organise time for this to happen.

Planning your work with the teacher

The teacher is responsible for planning the mathematics curriculum of the children. Because you have a close involvement with individual children and their learning, it is important that the teacher includes you early on in their overall plans. You need to know what learning objectives the teacher has planned for the children so that you can help the children achieve them. You also need to know what learning you should look out for.

Assisting Numeracy provides examples of activities that you and the teacher may find useful to build into the children's work programme. There is help with what to look for in children's learning and at what age children are expected to achieve certain things.

At the back of this book, on page 95, we have provided a feedback sheet. This is intended to help with individual activities, rather than the curriculum as a whole. The teacher fills in the details of the activity and what the children might learn. You then fill in what happens during the activity and how well the children understood what they did. It is important to talk to the teacher as well as to fill in the form. In discussion, you can give valuable information to the teacher that you often would not write down. The feedback sheet gives you something to base the discussion on; it is not intended to replace your meeting with the teacher.

Assessing children

When you work on mathematics with small groups of children, they will talk to you about their ideas. Ask them to tell you more about these ideas and you learn a lot about how they understand mathematics and what they know. This is useful information for the teacher. It helps you and the teacher plan the children's next activities.

You can also watch children while the teacher is talking with the whole class to see if they are joining in and to assess how well they take part. You might offer some children quiet help during this time.

To assess children in this way, you and the teacher need to agree what mathematics you both expect them to learn and what the children might do or say to show they are learning it.

To the teacher

Teaching assistants in the school

Teaching assistants are an invaluable resource, and you need to plan carefully how to use them. As a school, you should formulate an agreed policy on how teaching assistants can benefit your children and their learning. Be clear about what you can reasonably expect the teaching assistants to do and what help and guidance they need. All schools should have a detailed job description for teaching assistants, and all teachers should know what their own role is in helping teaching assistants carry out their job. Make time for joint planning and feedback sessions.

Part of your school policy should contain guidance for the training of teaching assistants. *Assisting Numeracy* is a useful resource for training, and you can incorporate many of the ideas in this book into specific training objectives. A clear message is the importance of children developing independence in their learning. We have emphasised this by developing the role of the teaching assistant in asking children questions to further their thinking and help them solve problems for themselves. We have looked to teaching assistants to help develop children's learning processes, rather than concentrate on narrow skills and short-term outcomes of activities. This is an area where there is great scope for training. The class teacher has a clear role in providing mentoring support on a one-to-one basis.

Teaching, learning and planning

You need to be sure that teaching assistants share your objectives for the children's mathematical learning so that you are both working towards the same goal. It is vital to find time for the pair of you to have discussions early on about overall curriculum planning: give teaching assistants the opportunity to understand why you want children to experience certain activities and how the activities fit with your understanding of how children learn mathematics. Teaching assistants also need some sense of how your plans fit into the overall progress and development of children throughout primary school. It is not always obvious to those not trained as teachers what makes some experiences more significant than others in the classroom, and you need to be specific about what it is you want the children to learn.

The feedback sheet on page 95 offers an opportunity for dialogue with the teaching assistant on a short-term basis. Fill in the details of the activity you have planned with your teaching assistant and the learning objectives and intended outcomes. Your teaching assistant can then record what happens during the activity and what the children appear to learn. However, written feedback cannot replace in-depth discussions about children's needs and detailed assessment of children's learning.

Assessing children

Teaching assistants can give you valuable information about children's development in mathematics. In order for them to acquire this information and pass it on to you, they need to know what you intend the children to learn, and what to look for in children's speech and actions to indicate their knowledge and understanding. In *Assisting Numeracy*, we offer a repertoire of questions that a teaching assistant can ask during maths activities to probe children's thinking. Encourage the use of this style of questioning and foster the skills of listening to, and remembering, key features of children's responses. Make time to observe your teaching assistant working with the children. They may not realise how many skills they have developed, so offer them positive feedback.

You need to think about how best to involve your teaching assistant at the beginning and ending of a maths lesson, particularly when you are introducing mathematical concepts to children. Involve teaching assistants in discussions and ask them to join in your conclusions to the lesson. A valuable use of your teaching assistant's time is to focus on particular children and monitor their responses to the discussion. You need to agree beforehand what to look out for: for example, whether particular children appear to engage in the lesson, whether they make any attempt to respond to general questions, any specific responses they make, and so on. Your teaching assistant can also sit near children who need help and give them quiet and discreet support during these whole-class sessions. During a plenary session, your teaching assistant can help a particular child or group feed back to the class, perhaps displaying particular skills these children have acquired, such as making jumps on a number line or counting past 100.

Supporting children with particular needs

Teaching assistants sometimes work with certain children more than others and sometimes exclusively with individuals with particular educational needs. Plan the work for those children together with your teaching assistant and make sure you both monitor what the children are learning. It can be useful for the teaching assistant to move around the class in a supervisory role while you work with a small group, so that all children benefit from your teaching. We have included some activities for larger groups in *Assisting Numeracy* that the teaching assistant can use for this purpose, as well as games and activities suitable for working with small groups. The aim is for you and your teaching assistant to work as a team in the classroom.

How to use this book

Assisting Numeracy is divided into four sections:

Introduction

The first part of the introduction is for the teaching assistant to read. The second part is for the teacher or mathematics coordinator.

The introduction looks at current issues in mathematics teaching and the role of teaching assistants in schools today. It suggests ways in which to enhance the teaching assistant's role with proper planning and regular feedback sessions. It also offers advice for assistants who work with children with special educational needs or who are learning English as an additional language.

Basics

The second part of the book identifies key areas of mathematical learning, with an emphasis on number at primary school level. It guides you through the way in which children learn skills and strategies most effectively. Used along with the 'Useful information' section, it will help you get a feel for the basics of mathematical thinking. There are also lots of tips on how to help children grasp these basics and get a good understanding of how and why we use mathematics.

Activities

The third section looks at the mathematics in ordinary classroom situations: story-telling, songs and rhymes, playing in the home corner or the playground and going on school outings. It offers a range of games and activities to help you make the most of all these situations.

Useful information

This is a comprehensive reference section offering information on

• mathematical equipment
• a glossary of mathematical words and their meanings
• a list of what children in each school year should be able to achieve in mathematics
• further resources that might be of interest to you.

2 Basics

Children need to express their ideas in order to understand fully what they are doing.

Language and mathematics

The importance of talking

Children need to talk about the mathematics they are learning. Talking helps children make their ideas clearer to themselves, and it makes them think about the mathematics they are doing. They get used to using new words and understanding what they mean.

When children talk, they also share ideas. They can learn from each other. They also sometimes disagree with each other about ideas. When they argue and explain, they gain more understanding.

In one school, a group of children were asked to add 6 + 18 and then say how they worked it out. One child explained: "I didn't know the answer, so I turned it round and put 18 in my head and used my fingers to count on 6." The adult asked her why she turned it round. She said: "Because when I put the big number first, I have enough fingers left to count on 6."

The other children joined in the discussion, talking about how to do calculations that are 'too big'. They offered ideas about always putting the bigger number first and about looking for numbers that make 10, such as 8 and 2. Some children in the group learned new ideas from this discussion.

Because you work mainly with small groups of children, you can encourage them to talk. Ask questions and listen to the replies and encourage the other children to listen, too.

The importance of listening

When you listen to children talking about mathematics, you can find out a lot about what they do, and don't, understand. You can get them to think things through carefully by asking them to explain more clearly what they mean. Don't say, "That's wrong", but ask, "What do you mean? How did you work it out? Does everyone else agree?"

Sometimes, listen without saying anything, but make a note of what children say, including errors and misunderstandings. You can then discuss these later with the teacher and plan how to do more work on them. Don't expect to be able to correct misunderstandings in a single lesson.

Listening gives you a chance to analyse what children are thinking. One child did the calculation 45 − 9 in his head. He explained his method in the following way: "I split up the 9 in my head into 5 and 4. 45 take away 5 leaves 40. Then 40 take away 4 leaves 36."

This explanation tells you a lot about how much the child understands. He clearly knows about how numbers work. Other children could learn from his expertise, as well as sharing their own ideas about how to make a calculation easier.

The importance of asking questions

It is better to ask children questions than to tell them the answers. By asking questions you encourage children to think. By telling them the answers you do their thinking for them. It is useful to have a variety of questions in mind to ask children in different mathematical situations. In the course of the school year, you will find these come to you automatically.

Let's look at an example of an activity and see the range of questions you might have in mind for that activity:

On a number line, put circles round the multiples of 4.

The first problem might be that a child does not recognise the word 'multiple'. Instead of explaining straight away, try to find out a bit more about what they do know:

Does the word 'multiple' remind you of another word?

What about 'multiply'? What does that mean?

This may encourage the child to think through what the problem is likely to mean. If not, you will need to explain what a multiple is. You may then need to prompt the child to think more about what to do:

How could you find out the multiples of 4?

Now you know what the multiples of 4 are. What do you have to do next?

If the child gets stuck during the activity, ask questions to help them carry on:

Have you ever done anything like this before?

Can you tell me what you have done so far?

Would it help if you used counters? Or a 100-grid?

When the child has finished, you can ask them to check their work, even if it is correct:

Are those all multiples of 4? How do you know?

Would you know if you had made a mistake? How would you know?

Shall I put a circle round 25? Is that a multiple of 4? Why not?

Closed and open questions

You can ask children two different types of question: closed questions and open questions.

Closed questions

A closed question has an answer that is either right or wrong.

What must I add to 3 to make 10?

There is just one correct answer: 7.

You can use closed questions to find out what facts a child knows. This can be useful. For example, sometimes you need to find out if a child knows the answer to a question such as: "What are three fives?"

Open questions

An open question is a question that can have several different answers. The answers can all be correct or else perfectly acceptable in the context.

What numbers add together to make 10?

What does 'equal' mean?

The first question has several correct answers. The second question doesn't have particular answers in the same way: it is really a way of getting children to explain. You may not understand their explanation, in which case you can say, "I don't understand. Can you try and explain again in a different way?" But try to avoid saying, "That's wrong."

When you use open questions, you encourage children to think things through. When you ask an open question such as: 'What numbers add together to make 10?', children can think of more than one solution. They can gain ideas and inspiration from each other and often become interested in patterns of numbers.

$3 + 7, 4 + 6, 5 + 5 \ldots$

$2 + 8, 2 + 2 + 6, 2 + 2 + 2 + 4 \ldots$

$9\frac{1}{2} + \frac{1}{2}, 8\frac{1}{2} + 1\frac{1}{2}, 7\frac{1}{2} + 2\frac{1}{2} \ldots$

This is an excellent way to stimulate children into thinking about numbers. It gets them thinking much more than if you ask a closed question like 'What is $3 + 7$?'

Here are some useful questions to ask children while they are working on maths activities. They will encourage them to think about and explain what they are doing.

What are you doing? How are you doing it?

What do you know so far?

What do you need to find out? How could you find it out?

Have you done anything like this before?

Does this remind you of anything?

What does that mean?

Changing a closed question to an open question

Sometimes, it is easier to think of a closed question than to think of an open one. However, most closed questions can be changed into open questions. Here are some examples:

Closed questions	Open questions
Does that symbol mean add or subtract?	*What could that symbol mean?*
Does that fact help you?	*How does that fact help you?*
Can you count the buttons?	*How could you count the buttons?*
What do 4 and 2 and 7 add up to?	*What numbers can you make using 4, 2 and 7?*
What is 6×5?	*If 6×5 is 30, what else can you work out from that fact?*
Is 24 an even number?	*What even numbers are there greater than 10?*

 Open questions can encourage children to think more than closed ones do.

Mathematical vocabulary

Children gradually build up a knowledge of mathematical words and their meanings. They should be able to understand these words when others use them, and they should be able to use appropriate vocabulary themselves. The teacher may include specific mathematical words in the lesson plans. The feedback sheet on page 95, for you and the teacher to use, includes a space for special vocabulary: 'Words to use'. The glossary at the back of the book explains some mathematical words.

Give children enough time to think of the answer.

Pay attention to children's use of mathematical language. When you are aware that children are not using a useful mathematical word, find the right way to drop it into the conversation. You can do this by repeating what they say, but adding the new word. For example: "You are telling us you have made that modelling dough into a square shape, with squares all the way round. So you have made a cube. Can anyone else make a cube?"

Sometimes, everyday words have a special mathematical meaning. We say, "Just half a cup for me", but we mean 'not as full as usual' rather than 'exactly half'. In mathematics, we expect 'half' to mean precisely 'half'. The word 'difference' also has a special meaning in mathematics. 'The difference between 3 and 5' means comparing the numbers 3 and 5 and looking for a numerical difference. The difference between 3 and 5 is 2 (because 5 is 2 more than 3). We need to explain to children that some words have a special meaning in mathematics. Only then can we expect children to use words correctly and accurately.

Useful tips

- You can learn a lot about what children do, and don't, understand by listening to them talk.
- Ask children to explain things to each other and listen to what they say. Children learn from each other, and you can find out what they know and understand.
- Keep a notebook with you when you work with children. Make a note of things children say that show what they do understand and what they don't understand. Pass on this information to the teacher.
- When you ask children questions, give them time to reply. Try counting to 5 in your head before you ask another question. Children need enough time to think.
- Encourage children to answer questions with a whole sentence.
- Ask the teacher if there is any special vocabulary you should emphasise over the next two or three weeks.
- Expect children to use the correct mathematical words. Introduce the correct words if the children don't use them.
- Try not to tell children the answers. Ask them questions instead.

When English is an additional language

Children who are learning English will already understand all sorts of things about mathematics. However, it is not easy to find out what they know if they cannot tell you. You may find it useful to have some simple number games and activities up your sleeve that children can learn by watching other children, rather than by listening to your explanations. Children who are learning English can play or work with children who already understand what to do. You can then observe them to see what they do seem to know about numbers.

When children are learning English, you want them to increase their vocabulary, and you want to get them to talk. Where appropriate, encourage children to talk in complete sentences, but allow one- or two-word answers if they still lack confidence. Games involving taking turns are a useful way of learning English and mathematics together. Make sure that the game is easy to play and that children have to say something simple such as the next number on the card. They can watch what other players say and do and join in when they understand what to do.

Sometimes, children who are new to the school, and have been taught in a different way, take a while to get used to a new way of learning maths. It will take longer if they are also learning to speak English. Some children may be familiar with another number script, and will be learning a new set of symbols. You need to bear all this in mind when working with these children, so give them time and the opportunity to show you what they do know.

The following game using number cards is an example of an activity that requires the children to say numbers out loud.

Years 2 to 4

Tricksy

A game for 3 or more children

You need:

- 0–100 number cards (the pack need not be complete)

The game

- Shuffle the cards and deal six to each player. Players may look at their cards.
- The first player chooses one of their cards, puts it in the middle of the table and says the number.
- The other players choose a card in turn, put it on the table and say the number.
- The player who puts down the highest number wins the trick. (A trick is the set of cards played in that round.)
- Keep playing until all the cards are on the table.
- The player who has won most tricks after several rounds is the overall winner.

Easier

Use 0–50 number cards or two sets of 0–20 number cards.

Harder

Children look at their cards and decide how many tricks they think they will win. They write down their guess and see if they can make it come true.

Things to say

What is that number called?

Which of those two numbers is greater?

Can you show me that number on a number line?

Using 100-grids with language learners

100-grids show the numbers from 1 to 100. You can make good use of them with children who are learning English. Children can point to the numbers while saying the numbers in their own language; the English speakers can listen and try to imitate and learn the number words. Help the English speakers notice any sound patterns such as we have with numbers above 20 (sixty-one, sixty-two, sixty-three ...).

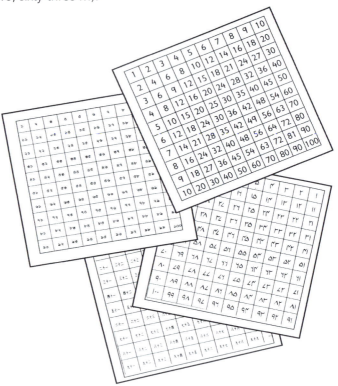

You may be able to find a 100-grid in a different script (Bengali, Chinese, Arabic, and so on). If so, you can compare the way the numbers look on this grid and your usual grid. Find the number 44 on each grid and see if it is in the same place and make sure it really is 44. Children who know a particular script can teach the other children about it.

Cut a 100-grid (in the usual script or one of the other scripts) into three or four pieces so that the children can put them back together again by sequencing the numbers correctly. Older children may cope with more pieces.

Mental mathematics

Mathematics is a mental activity
– it takes place in the head.

Mathematics is a mental activity and therefore can seem daunting and difficult. Often, when there is too much to keep in your head at once, you will write things down to keep track of what you are thinking. For example, the stallholder at the market lists the prices of what he is selling you on a paper bag, then adds it all up. Sometimes, he can do the calculation in his head, and if you want to check his arithmetic, you have to work it out in your head at the same time. Then you both have to work out the correct change.

Writing down numbers can help you with your calculations because you don't have to remember so much, but the actual mathematics takes place in your mind.

What are three nines?

I know three tens are 30.

9 is one less than 10. So three nines will be three less than 30. It's 27!

Children need to learn lots of mental methods of calculating. If we talk to children about the different methods, we can help them choose the ones that make working things out quicker and easier. Children can compare ways of working with numbers to see which ways are easiest and most appropriate. Here are some examples of what children should learn in mental arithmetic:

- Remembering number facts like $3 + 7$ or 3×5 or $24 \div 6$
- Recalling these number facts quickly
- Ways of working things out in their heads, given a bit of thinking time: for example, $72 - 35$, 24×3 or 10% of £345
- Solving problems like 'Can I buy three 26p stamps for £1?'

$$72 - 35$$

People have different ways of working out a calculation like $72 - 35$. Some methods work well, and some are quicker than others. Children need to talk about and compare different methods. They can learn quicker methods from you and from each other. For example, it is possible to take away 35 one at a time from 72, using your fingers, but this method is slow and unreliable. Here are some other methods that work better, with number line pictures to help describe them:

- Take 2 from 72 to make 70, take 30 from 70 to leave 40. That's 32 I have taken away. Take away 5 and add on 2 is the same as taking away 3 from 40. It's 37.

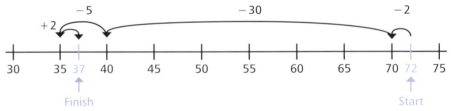

- Take 30 from 72 to make 42. Take 2 from 42, that's 40. Three more left to take away, that's 37.

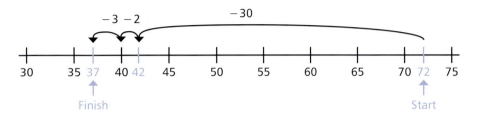

- Add on 5 from 35 to 40, add on 32 from 40 to 72, add the 5 and the 32 to make 37.

$$35 + 37 = 72$$
$$\text{so } 72 - 35 = 37$$

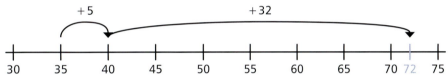

All these methods are good for working out $72 - 35$. But to use them, you need to know some important things:

- It helps if you know simple number bonds like 7 take away 3. This makes it easy working with tens like 70 take away 30.
- Being able to take small numbers away from 10 makes it easier to take numbers away from a tens number like 30 or 70.
- It helps if you know doubles of numbers.
- It helps if you break up a calculation into smaller parts.
- You can use addition to solve a subtraction problem.

These are examples of the important things children need to learn about numbers and how they work. Teachers will plan for children to learn particular methods. Talking with children will help them think about these methods and learn from you and from each other.

Writing things down

Children learn written methods in order to do calculations that are too long or complicated to do in their heads. Written methods are best when working mentally is hard to do. But if children are good at working mentally, they can save time and effort by working in their heads. The school will have its own favourite written methods; make sure you know what these are so that you can support children as necessary.

While children are using mental methods, they can jot down numbers as they work to help them remember, like taking notes. Encourage children to find helpful ways of using scrap paper and pencil to make these jottings. This informal written work is a helpful record of how children are getting on and tells you a lot about how they worked things out.

When children write down a calculation that they can work out mentally, encourage them to write it horizontally, on the same line.

$106 - 8 =$

This allows them to work it out in their own way. If children write a calculation vertically, down the page, they may then use a longer written method even if they could do it easily in their heads.

$$\overset{9}{\cancel{1}}\overset{1}{0}6$$
$$\underline{-\ 8}$$
$$98$$

When children write down calculations that they cannot work out in their heads, they need to record these neatly in a way that will help them work them out:

346		346
988	not	988
$+465$		$+465$

What are four 19s?

I know four 20s are 80.

19 is one less than 20.

So four 19s will be 4 less than 80.

It's 76!

- Mathematics is something we do mentally. Encourage children to work things out in their heads.
- Help children learn their number facts and times tables.
- Get children to discuss their mental methods with each other and with you. Ask them which methods are the quickest and easiest.
- When they are doing calculations in their heads, children don't need to write anything down unless they are practising or they are jotting things down to keep track of the calculation.
- Encourage children to work things out in their heads, only using fingers, number lines or counters when necessary.
- There is no 'right' way of working something out in your head. If the method works sensibly, then it's OK.
- Give children practical problems to calculate, such as: "Dinner time is at 12:15, and we have got 20 minutes to go. What time is it now?"; "Only three children are allowed in the home corner and there are five in there. How many have to come out?"
- When a child's calculation is too hard to do in their head, they need to write them down neatly in the way they have been taught.

Things to say

You have both got different answers to that calculation.
Why don't you check your answers

How could you check that answer?

How did you work that out?

(Ask this question even when children have got the answer right. Children will notice if you only ask this when they get it wrong. And you do want to know how children worked out their calculations, right or wrong.)

Did anyone do it a different way?

Can you think of a story to go with those numbers?

(For example, a story to go with $3 \times 4 = 12$ might be: "Three sparrows laid four eggs each. How many eggs were there altogether?")

Mental strategies for children to use

Addition and subtraction

Start with the bigger number

When adding, start with the bigger number first.

> *You need to add 6 and 18. Start with 18 and add on 6.*

Use doubling

When adding a number to itself (doubling), children can often work out the answer easily: for example, 5 + 5 or 60 + 60.

You can use doubling when you want to add two numbers that are nearly doubles: just add or subtract the difference.

> *6 + 6 is 12, so what is 5 + 6?*
>
> *Two 25s are 50, so what is 27 + 25?*

To add 9, first add 10, then take off 1

If children can add 10 to a number, then they can add 9 by adding 10 and subtracting 1.

Similarly, if children can subtract 10 from a number, then they can subtract 9 by taking off 10 and adding back 1.

You can add and subtract 11 easily, too.

> *46 + 9? Try 46 + 10, then take away 1.*
>
> *What is 27 − 9? Do 27 − 10. Do we need to add 1 back on or take another 1 away? Why?*
>
> *36 plus 11? Do 36 + 10, then add another 1.*

Do a bit at a time

> *You want to subtract 32 from 59? First subtract the 30, which leaves 29. Now take off two ones, that's 27.*
>
> *You want to add 17 and 6? 6 is made up of 3 + 3. 17 + 3 is 20. Add the other 3 … yes, 23!*

Fill the grid

An activity about adding and subtracting mentally
For 2 children or 4 in pairs

You need:

- Coloured pens and pencils
- Set of 0–10 number cards
- Set of 0–20 number cards
- Blank 3 × 4 grid to share

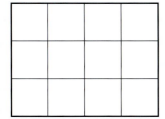

The activity

- Each child chooses a different-coloured pen.
- Shuffle each set of cards and place them face down on the table in two separate piles.
- Children take turns to pick a card from each pile. They add or subtract the two numbers and write the answer in any space on the grid. They then return the cards to the bottom of either pile.

  They carry on until the grid is filled with numbers. You may need to agree with the children whether they can write the same number in more than one square.

- Next, children take turns to pick a card from each pile and add or subtract the two numbers. If the answer is on the grid, they cross it out with their pen; if not, they do nothing.
- Children try to cross out five squares as quickly as possible.

The aims

- Practising number bonds (see Glossary, page 87)
- Working out calculations for numbers to 30

Adapting the activity for different ages

Younger children

Children use a 2 × 3 grid and two dice numbered to 3 or 6.

Older children

Children use a 4 × 4 grid. They pick two cards and roll a dice so that they have three numbers to work with. Allow addition, subtraction or multiplication.

Things to say

Which of those numbers will you start with?

Which will give you a bigger answer: adding or subtracting? Try it and see.

Ypu need a 4. How could you get 4 by adding two numbers? How could you get 4 by subtracting two numbers?

Things to notice

Can the children:

- add and subtract, using any pair of numbers to 10, such as 4 + 5, 3 + 6 or 9 − 5?
- start with the larger number when adding or subtracting?
- begin to use some useful strategies to do mental calculations (see page 28)?
- explain how they did a calculation?

Mental strategies for children to use

Multiplication and division

Think about what the answer should look like

When multiplying, children can check their answer by using what they know of the tables.

All answers in the 2×, 4×, 6× and 8× table are even.

All answers in the 5× table end in 5 or 0.

All answers in the 10× table end in 0.

Use other checks, too

Halving an odd number gives you a remainder; halving an even number doesn't.

Work it out from a fact you know

To work out six threes …

I know that five threes are 15. Add one more 3 and I get 18.

Use doubling

Multiplying by 4 or 8 is easy if children know how to double.

You need to know 23 × 4. Double 23, then double again.

To do 12 × 8, start with 12 and double it, then double again, and again.

Use halving

Dividing by 4 or 8 is easy if children know how to halve.

You need to know 124 ÷ 4. You can halve 124, then halve again.

To do 64 ÷ 8, start with 64 and halve it, then halve again, and again.

How many sticks?

An activity about multiplication
For 3 or 4 children

You need:

- Set of 16 shape cards (make four of each type)
- 1–6 dice
- Headless matches

Draw the shapes as if they were made from sticks.
All the sides should be the same length.

The activity

- Place the shape cards on the table, face down.

  Children take turns to pick a card and roll the dice.

  The card shows what shape each child needs to make, and the dice shows how many of the shapes they need to make. Children work out the number of sticks they need. They count out that many sticks, then check by making shapes on the table.

How many sticks will you need to make five triangles?

The aims

- Building up children's speed, knowledge and confidence
- Practising multiplication number facts to 6×6

Adapting the activity for different ages

Younger children

Children use only triangles (three sides) and pentagons (five sides).

Older children

Children don't make the shapes with sticks but just say the number. The other children check whether it is the correct number. (They could use a calculator to check.)

or

Use 1–10 number cards instead of dice.

or

Use 1–36 number cards, which signify a number of sticks, and the shape cards. Children take a shape card and a number card. They read the number, imagine they have that number of sticks and say how many of that shape they could make, and how many sticks (if any) would be left over.

Things to say

How can you work that out quickly?

Let's all count in fives together to help.

Can you count out the sticks you need in twos?

If I give you 12 sticks, how many triangles could you make?

Things to notice

Can the children:

- count in threes, fours, fives and sixes?
- explain how they know they have made a mistake? For example: Do they realise that 6 times 3 cannot be 17 because all the answers in the 6 times table are even?
- use facts they already know to help them work out new ones? For example: "I know 5 times 3 is 15, so 6 times 3 must be 15 add 3."

Children need lots of practice before they get a feel for different measurements.

Measures

In this section, we look at the following measures, all of which children need to learn about in primary school:

- length
- weight
- volume and capacity
- time
- money

There is a practical activity for each of these measures. All the measures, except time, use metric measurements: that is, they are based on 10 and its multiples.

We get better at measuring the more we do it. As adults, most of us have a clear idea of what a pint is and how long a metre is. We can do rough measurements without using measuring tools. We can tell by lifting a kettle whether it has enough water in it and can shake the right amount of coffee into the cup to make it the right strength. We know how big a space we need to park the car and whether we need one or two shopping bags for all the goods piled up at the supermarket checkout.

Children, however, need lots of practice before they understand measuring and have enough experience about different measurements. Young children misjudge how much water to put in a jug. They are certain they need a huge piece of paper to cut out a small shape. They pick a small piece of paper to wrap a big box.

At school, children need plenty of opportunities to measure things properly with measuring tools so that they get used to different metric measures. Younger children also need practical tasks such as mixing paint and cutting paper to get a feel for measuring things informally.

Length

Basics

There are 100 centimetres in a metre.

A metre is a little over a yard – about the distance from an adult's chin to the tip of the index finger, with the arm fully extended.

A kilometre is 1000 metres; a distance of 8 kilometres is roughly 5 miles.

Vocabulary to use

long, short, tall, high, low, wide, narrow, deep, shallow, thick, thin, far, near, close, longer, shorter, longest, shortest, metre, centimetre, millimetre, approximately, estimate, about, roughly

Useful tips

- Make sure children measure from the 0 on tape measures and rulers and don't count the extra bit before the 0 mark.
- If the tape/ruler has centimetres on one side and inches on the other, make sure children know which is which and use the correct side.
- Talk about how much the 'little bits' between the numbers are worth.

How much is each of those little distances?

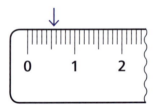

- Encourage children to record their findings. They could make a simple table.
- An estimate is a considered guess. Let children compare their estimates with their measured results as this will help them estimate more accurately in the future.

This is one way of recording the activity opposite.

	nearest wall	furthest wall	comments
estimate	5 m	17 m	
measured	$3\frac{1}{2}$ m	25 m	

What's the distance?

An activity about measuring length
For 2 or 3 children

You need:

- Chalk
- Pencil and paper
- Metre sticks

The activity

Say to the children:

- Stand anywhere in the hall (or playground) and mark where your feet are.
- Estimate the distance from where you are standing to the nearest wall. Write down your estimate.
- Now measure the actual distance, using a metre rule, and write it down.
- Go back to your chalk mark. Estimate the distance from there to the furthest wall. Write down your estimate.
- Now measure the actual distance, using a metre rule, and write it down.
- Stand what you think is about 3 metres from one of the walls.
- Find out how good your estimate of 3 metres is by measuring the actual distance.

The mathematics children practise

- Making sensible estimates
- Using a metre rule
- Recording estimates and measurements
- Measuring length with some accuracy
- Getting a feel for how long a metre is

Adapting the activity for different ages

Younger children

Children count strides (large steps) instead of using metre sticks.

Older children

Children use a surveyor's tape instead of metre sticks to measure longer distances. Most surveyor's tapes are measured in tenths of a metre.

Things to say

What does 'estimate' mean? Will a wild guess do?

Can you see an imaginary metre rule in your head? Can you imagine measuring from here to the wall with it? Does that help you estimate?

Now you have measured about half of the distance, would you like to change your estimate?

You say the distance is four metres and a bit. Can you be more accurate about how much the 'bit' is? About half a metre? A quarter of a metre? How many centimetres are there in a metre?

Things to notice

Can the children:

- use what they know about metres to help them estimate?
- mark carefully on the floor where the end of the metre stick comes each time or do they work haphazardly?
- suggest ways of measuring the 'bit' when the distance is not exactly whole metres? Do they, for example, say, "It is half a metre" or "It is about 10 cm"?
- use a metre stick to measure centimetres accurately or do they count from the wrong end?
- make more accurate estimates with practice?

Weight

Basics

The units of weight used in schools are grams (shortened to g) and kilograms (shortened to kg).

There are 1000 grams in a kilogram.

There are about 30 grams in an ounce, and a kilogram is just over two pounds.

Stones, pounds and ounces, known as imperial measures, are being phased out, but children do need to know something about these units, too.

Vocabulary to use

heavy, light, heavier, lighter, heaviest, lightest, weighs, balances, gram (g), kilogram (kg), estimate, roughly, nearly, approximately, about the same as, in between

Useful tips

- If using kitchen-type scales with a dial, make sure the pointer is set at 0 when the pan is empty.
- Check that children know what units they are using: grams, kilograms, ounces, and so on.
- Talk about how much the 'little bits' between the numbers are worth.

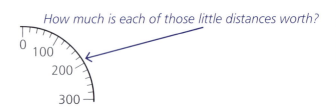

How much is each of those little distances worth?

- Encourage the children to record their findings. They could make a simple table.
- An estimate is a considered guess. Let children compare their estimates with their weighed results as this will help them estimate more accurately in the future.

This is one way of recording the activity opposite.

This is the order we think they are in. We think the water is heaviest and the flour is lightest.

	heaviest				lightest
estimate	water	sand	soil	rice	flour
weighed					

Find the heaviest

An activity about weighing
For 2 or 3 children

You need:

- Three same-size yoghurt pots
- Materials to fill the pots: water, sand, rice, pebbles …
- Weighing balance

The activity

- Children take a pot each and fill it, each choosing a different material. (You can fill the third pot if there are only two children.)
- Children work as a group to put the pots in order (heaviest first), estimating only at this stage. In order to do this, they need to compare two pots at a time by holding one in each hand, then another two, and so on. When they have finished, children record their estimates in some way.
- Children then use the balance to compare the pots and put them in order.

The mathematics children practise

- Making sensible estimates
- Using weighing instruments
- Recording estimates and measurements

Adapting the activity for different ages

Younger children
Children use only two pots.

Older children
Children use five pots.

Children use the balance and a selection of weights: 5 g, 10 g, 20 g, 100 g, 500 g … They find out the weight of each pot in grams, then put them in order.

Children use kitchen-type scales to find out the weight of each pot in grams, then put them in order.

Things to say

What do people use to weigh things?

I wonder if all the pots weigh the same. What do you think?

What is meant by 'estimating'? Does it just mean guessing? (An estimate is a considered guess: for example, a guess made after picking up the pots and noticing how heavy they feel.)

Which do you think is the heaviest of all the pots? And which do you think is the lightest? Why do you think that?

How good was your estimating? Do you think you would be more accurate another time?

Things to notice

Can the children:

- use a balance to compare the weights of three or more objects and put these in order?
- use a balance and weights to weigh an object in grams?
- read the dial on the kitchen scales? Do they understand the intervals between the numbers?

Volume and capacity

Basics

Liquid volumes are measured in litres (shortened to l) and millilitres (shortened to ml).

Solid volumes are measured in cubic centimetres (shortened to cc).

A centicube, for example, has a volume of 1 cc (or 1 cm³).
One ml of water takes up the same space as 1 centicube.

There are 1000 millilitres in a litre.

A litre is just over two pints.

Pints and gallons, known as imperial measures, are being phased out, but children do need to know something about these units, too.

Vocabulary to use

full, empty, brim, overflowing, holds, holds most, holds least, amount, refill, litre, half a litre, millilitre, funnel, container, cylinder, estimate, roughly, approximately, nearly, about the same as

Useful tips

- Keep children's work away from water.
  Don't use sand and water in the same lesson!
- Check that children know what units they are using: millilitres, litres, pints, fluid ounces, and so on.
- Help children practise how to read calibrations on jugs and cylinders.

  How much water does that mark represent? ——————→

  —— 100 ml

  How much is each of those little distances? ——————→

- Encourage the children to record their findings. They could make a simple table.
- An estimate is a considered guess. Let children compare their estimates with their measured results as this will help them estimate more accurately in the future.

This is one way of recording the activity opposite.

	estimate	measured	comments
more than 1 litre	Shampoo bottle	yes	
about 1 litre	yoghurt pot	no	

Hunt the container

An activity about measuring volume
For 2 or 4 children

You need:

- Slips of paper marked:

more than 1 litre

about 2 litres

more than $\frac{1}{2}$ litre, but less than 1 litre

less than $\frac{1}{2}$ litre

about $1\frac{1}{2}$ litres

more than 1 litre, but less than 2 litres

about 500 millilitres

- A selection of plastic bottles, jars and bowls that hold between $\frac{1}{4}$ litre (250 ml) and 2 l. You may want to label them A, B, C, and so on.
  - Materials that can be poured: water, sand, rice …
  - Measuring jugs and cylinders: 500 ml, 1000 ml …

The activity

- Deal out the slips of paper so that each child has one or more.
- Each child or pair of children has to select a container which they estimate will hold roughly the amount written on their slip of paper.
- Children fill their container with sand or water. They pour the contents into a measuring jug, measure the quantity and record their results.

The mathematics children practise

- Making sensible estimates
- Using calibrated containers for measuring capacity
- Reading capacity scales
- Recording estimates and measurements

Adapting the activity for different ages

Younger children

Children stick to finding containers that hold either 'more than 1 litre' or 'less than 1 litre'.

Older children

Children use slips of paper with more difficult specifications such as:

$\frac{1}{2}$ litre to 750 ml

10 ml to $\frac{1}{4}$ litre

50 ml to 100 ml

less than 5 ml

between 250 ml and 500 ml

Things to say

I wonder whether that jug can hold more than one litre or less.

What can you use to help you make a good estimate? (Encourage the children to look at a measuring jug to get a feel for how much a litre is.)

Which is the most sensible measuring jug to use for finding the capacity of that jar?

Is that more or less than one litre? How do you know?

Are you surprised by any of the results?

Things to notice

Can the children:

- recognise when a tall, thin container holds less than a short, squat one?
- compare the capacity of two containers by pouring from one to the other? Do they take notice of overflowing? Do they notice space left at the top?
- make sensible estimates when choosing jugs to use for measuring?
- work carefully and accurately?
- read the calibrations in the measuring jug?

Time

Basics

There are 60 seconds in one minute.

There are 60 minutes in one hour.

There are 24 hours in one day.

An analogue clock is one with rotating hands.

A digital clock displays the time in digits.

Vocabulary to use

clock, watch, o'clock, face, digital, analogue, minute hand, hour hand, early, late, earlier, later, earliest, latest, fast, slow, midnight, midday, noon, hour, minute, second

Useful tips

- Practise reading the time with children whenever the opportunity occurs, using both school clocks and wrist watches.
- Talk about the time of day when things happen:

  Lunch is at half past twelve.

  Can you show me where the hands will be at half past twelve?

- Talk about time passing:

  There are five minutes left before playtime. Time to pack up!

Find the pairs

A game about reading the time
For 3 or 5 children

You need:

- 15 cards showing random times in digital form (stick to multiples of 5 minutes)

- 15 cards showing the same times in analogue (clock face) form. You could make these cards using a clock stamp.

- A real digital timepiece such as a child's watch and a real analogue timepiece such as a school clock (optional)

The game

- Put the digital cards in a pile on the table, face down. Spread out the analogue cards on the table, face up.
- Each child takes a card in turn from the pile of digital cards, says the time shown and finds the matching analogue card from the table. The children hold onto their pairs.
- If a child makes a mistake, the digital card goes back to the bottom of the pile and the analogue card remains face up on the table.
- When the children have finished, they play again, this time putting the analogue cards in a pile and spreading the digital cards face up.

The mathematics children practise

- Counting in fives
- Reading the time in five-minute intervals on digital clocks
- Telling the time in five-minute intervals on analogue clocks
- Recognising equivalent forms of the same time

Adapting the activity for different ages

Younger children
Children stick to simple times: o'clock, half past, quarter to, quarter past.

Older children
Children use cards showing 'harder' times such as 7:21 pm.

Children use cards showing times on the 24-hour clock, such as 19:21.

Things to say

Look at this watch. Which hand shows the minutes? Which hand shows the hours?

Look at this digital watch. Do you know which number shows the minutes? And which shows the hours?

How many minutes after 2 o'clock is quarter past 2? You can use a real clock to help you work it out.

What time does this card show? What do you normally do at that time of day?

Things to notice

Can the children:

- work out the analogue time by counting in fives, using a clock face?
- work out digital time, especially between half past the hour and the hour: for example, 7:35?
- understand that 'five past' is written with a zero before the five: for example, 9:05?

Money

Basics

100 pence is £1.

Vocabulary to use

penny, pound, how many?, how many more?, how many fewer?, how much?, cost, price

Useful tips

- Allow children to handle real coins whenever possible: for example, when shopping or handling dinner money.
- Use real coins when playing games. Counting how many there are at the start and checking at the end is a valuable mathematics activity in itself.
- Some children are good at calculating with money. Use this to help them with other mental maths:

  You need to add 23 and 36. Imagine that's 23p and 36p. Can you add up those amounts?

Reception and Year 1

Make it up to …

An activity about adding coins
For 3 or 5 children

You need:

- Money dice showing 1p, 1p, 2p, 2p, 3p, 3p
- A saucer of 1p coins

The activity

Say to the children:

- Agree who will start. That person rolls the dice.
- What does the dice say? Can you read out the number of pennies it says?
- Now take that number of pennies from the saucer.
- Say how many more pennies you need to make it up to 5. Take that number of pennies, too.
- Check whether you were right.
- Now it is the next person's turn.

The mathematics children practise

- Familiarity with coins
- Adding numbers mentally to make 5

Adapting the activity for different ages

Younger children

Children each have a board marked with 10 or 12 pennies. They roll a dice with 1, 2 or 3 spots to collect pennies and fill up the board.

Older children

Children use 2p as well as 1p pieces and the same dice and say how many pence they need to make the amount up to 10p.

Children use a dice showing 1p, 2p, 3p, 4p, 5p, 6p and say how many pence they need to make the amount up to 20p.

Things to say

How did you work that out?

I wonder if you can show me 5 with your fingers.

Can you show me on your fingers how many more pennies you need?

Things to notice

Can the children:

- say how many more pennies they need to make 5?
- say how many more pennies they need to make 10?
- use their fingers to work out how many more pennies they need to make 5?
- use their fingers to work out how many more pennies they need to make 10?

3 Activities

Maths games and activities

Games and playful activities help children with learning and encourage them to talk to each other about what they are learning. They are also the perfect opportunity for children to take control and feel in charge.

Games for small groups

There are games in this section for each age group. The games are simple, and you can adapt them for younger and older children. Talk to the teacher about the learning objectives for each game.

The games are supported by:

- questions to ask children to make them think about the mathematics
- things to notice about what children are learning
  (You can make a note of these and tell the teacher about progress children make, and any problems they have.)
- ways to adapt the activity for younger and older children
- information on the mathematics the children practise by playing these games

Useful tips

- Build up a collection of activities to play with children at odd moments, such as 'I spy numbers' or number games involving a dice (keep one in your pocket).

- Collect together the things you need for each game. Put them in a zipped plastic wallet with a list of contents. Children can check that nothing is missing at the end of the game.

- Children can learn the game from each other: teach one group how to play a game, then ask a reliable child to teach the game to another group.

- Children can suggest ideas for changing the games to make them harder or easier or more interesting. Make a note of their ideas. Try them out and encourage the children to discuss them.

- Discuss with the children the mathematics they have been practising. This helps children apply their newly learned skills in other mathematical contexts.

Give-away

A game about counting to 10
For 3 or 4 children

You need:

- 1–6 dice (spots or numerals)
- A saucer or tub for each child, each containing 10 counters

The game

- Children take turns to roll the dice, say the number and give that number of counters to the child on their right. If they don't have enough counters to do this, they miss that go.
- The player on the left takes the next turn.
- The first player to get rid of all their counters is the winner.

The mathematics children practise

- Counting to 10
- Recognising the arrangement of spots or numbers on a dice

Adapting the activity for different ages

Younger children
Children use a dice marked 1, 1, 2, 2, 3, 3 and six counters each.

Older children
Children use two dice and add the numbers together. They need 20 counters each.

Things to say

What number did you get?

How many counters must you give to Sasha?

I wonder how many counters you have got now. Can you count them?

If Joe gives you five counters, how many will you have? And how many will Joe have left?

Things to notice

Can the children:

- recognise the pattern of spots on the dice without having to count? Recognise the numeral on the dice?
- give the correct number of counters to their neighbour?
- check that they are being given the correct number of counters?
- count how many counters they have left after each go? Tell their left from their right?

More than/less than

A game about ordering numbers
For 3 children

You need:

- 0–10 number cards
- Counters
- Dice made from a wooden or plastic cube, with three sides marked 'more than' and three sides marked 'less than', or a 'more than/less than' spinner

The game

- Put the number cards in a pile on the table, face down.
- Two of the players each pick a card, say the number and take that many counters.
- The third player rolls the dice and reads out the result: 'more than' or 'less than'.
- If it is 'less than', the player with fewer counters than their partner must say how many fewer they have. If it is 'more than', the player with more counters than their partner must say how many more they have.
- The player who gets to say the sentence about 'more' or 'less' gains a point if they say it correctly. Help them check by laying their counters in two lines, side by side, and counting the 'extras' in one line.

 *I've got 2 more than you.*

- The first player to score five points is the winner and rolls the dice in the next round.

Variation

Use 1–5 cards or 0–20 number cards.

Make 12

An activity about adding to 12
For 2 children or 4 in pairs

You need:

- Two sets of 0–10 number cards

The activity

- Shuffle all the cards together and share them out between the children. Each child lays out as many rows of cards as they can that add up to 20.

Variation

Children make agreed swaps with each other.

Take it

A game about adding numbers
For 2 to 3 children

You need:

- 2–12 number cards
- Two 1–6 dice

The game

- Children lay out the cards in order, face up. They take turns to roll both dice and add the numbers together. They take that number card (if it has not been taken already).
- The player who collects most cards is the winner.

Variation

Use 0–12 number cards and let the children add or subtract the two numbers.

or

Use a 10-sided dice and 0–20 number cards.

Number jigsaw

An activity about ordering numbers
For 1 to 3 children

You need:

- 0–30 number line or 100-grid, cut up into five or six pieces

The activity

- Children jumble up the pieces, then piece them together to reconstruct the number line or square.

Number books

An activity about recognising, reading and calculating with number
For any number of children

You need:

- Sheets of paper, plain or coloured

The activity

- Children choose three sheets of paper and stack these together. They staple or sew the pages in the middle and fold the sheets to make a little book.
- Children choose a number. This number is the theme of their book. They then fill the pages with information and images relating to that number.

  For example, they could:
  – look in magazines for printed versions of the number
  – collect examples of the number appearing in door numbers, telephone numbers or bus numbers
  – make up calculations that have that number as the answer.

Snake pit

A game about mental addition
For 1 to 3 children

You need:

• Pencil and paper

The game

Children begin with a score of 0 and keep a written record of their score as they go along. The aim is to reach 50 without falling into the snake pit.

These numbers are in the snake pit:

10   20   30   40

• Players take it in turns to roll two dice and choose one of the numbers to add to their current score. (You can help children check their calculation using a calculator after they have done it mentally.)

• If they land in the snake pit (that is, if their total is one of the snake-pit numbers), they must subtract 5 from their score.

• They then pass on the dice to the next player.

• The game ends when a player reaches or passes 50.

The mathematics children practise

• Mental addition
• Making choices

Adapting the activity for different ages

Younger children

Children start with 5 counters each. They use two dice, both marked 1, 1, 2, 2, 3, 3, and aim for 20.

The snake pit contains the numbers 5, 10, and 15. If a child lands in the pit, they lose a counter.

Older children

Children start with 100 and aim to reach 0. They roll three dice and choose two of the numbers to add together, then subtract the result from the total in the display. The snake pit contains the numbers 10, 20, 30, 40, 50, 60, 70, 80 and 90 and if a child lands in the pit, they add 10 to their score.

Things to say

Which of the numbers are you going to choose to add? Why are you picking that one?

What will you get if you add that number to your current score?

Can you say the answer without using your fingers?

I wonder what number you need to reach 50.

Things to notice

Can the children:

• work out what will happen if they choose either number, then choose the best number to avoid landing in the snake pit?

• add on numbers mentally?

• jot something down on paper to help them add?

Descriptions

A game about ordering numbers
For 3 to 6 children

You need:

- About 50 labels with brief descriptions of numbers such as 'ends in 0', 'even', 'in the 9× table'
- 0–50 number cards

The game

If there are more than three children, arrange them to play in pairs.

- Shuffle the labels and give nine to each player. Players arrange their labels in front of them.
- Shuffle the number cards and put them in a pile on the table, face down. This is the pick-up pile.
- The first player takes the top card and says the number. They decide whether it belongs with any of their labels. If it does, they place it on the label. If it doesn't, they put the card face up on a discard pile.
- The next player can take either the top card from the discard pile or the top card from the pick-up pile. They find a place for the card on their labels or discard it if they cannot put it anywhere.
- Players continue like this until one player has covered all nine of their labels. This player is the winner.

The mathematics children practise

- Thinking about whether a number is greater than or smaller than another number
- Learning odd and even numbers
- Learning multiplication tables

Adapting the activity for different ages

Younger children
Each child has just five of the easier labels.

Older children
Use 0–100 number cards.

Suggest children make up their own property labels to add to the set.

Things to say

Read out your labels one by one. Does that one apply to your number? What about that one?

Tell me some numbers that could go on that label. Are there lots of numbers that could go there or only a few?

Could that number card go on more than one label?

Which label will you choose to put that card on? Why?

Things to notice

Can the children:

- read a label and identify some numbers that could go on it?
- say which numbers are odd and which are even?
- say whether a number is greater than or smaller than another number?
- say whether a number is in a particular times table?

50 ideas for labels

ends in 0	in the 9× table	has a 2 in it	less than 40	more than 35
ends in 1	in the 8× table	has a 3 in it	less than 35	more than 30
ends in 2	in the 7× table	has a 4 in it	less than 30	more than 25
ends in 3	in the 6× table	has a 5 in it	less than 25	more than 20
ends in 4	in the 5× table	has a 6 in it	less than 20	more than 15
ends in 5	in the 4× table	has a 7 in it	less than 15	more than 10
ends in 6	in the 3× table	has a 8 in it	less than 10	more than 5
ends in 7	in the 2× table	has a 9 in it	less than 5	even number
ends in 8	has a 0 in it	less than 50	more than 45	odd number
ends in 9	has a 1 in it	less than 45	more than 40	odd and more than 2

Number squares

An activity about adding
For any number of children

You need:

- Set of 0–20 number cards for each child

The activity

- Each child chooses nine of their cards and arranges them in a 3 × 3 square.
- They then find the total for each column and row.

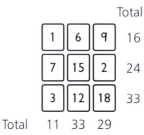

Total

1	6	9	16
7	15	2	24
3	12	18	33

Total   11   33   29

- Children rearrange the cards and find the totals again. How many different totals can they make?

Variation

Use cards with higher numbers.

Making numbers

An activity about adding and multiplying
For 2 children working together

You can split up a larger group into pairs, with each pair working on the activity at their own pace.

You need:

- Four 1–6 dice
- Pencil and paper
- 100-grid, felt-tipped pens

The activity

- Children roll all four dice and write down the numbers that come up.
- They use any or all of those numbers, in any way, to make as many different answers as they can. For example, with 1, 4, 4 and 6, a child could make:

$$14 + 46 = 60 \qquad 6 + 1 + 4 + 4 = 15 \qquad 6 + 1 = 7$$
$$4 + 4 = 8 \qquad 4 \times 6 = 24 \qquad \textbf{and so on}$$

- Children record their calculations on paper and cross off the numbers they make on a 100-grid, using felt-tipped pens. How many different numbers can they make this way?

Variation

Use cards with higher numbers.

Four in a line

An activity about adding and multiplying
For 1 or 2 children

You can split up a larger group into pairs,
each pair working on the activity together.

You need:

- Four 1–6 dice
- Pencil and paper
- Calculators or a tables chart (optional)
- 100-grid
- Counters

The activity

- Children roll all four dice and make calculations
  by adding and multiplying the dice numbers in
  any way they choose.

  For example, with 2, 3, 4 and 4, they might do:

  2 + 3 = 5   **(4 + 4) × 3 = 24** **(that is, 8 × 3 = 24)**

- They may use a calculator or a tables chart to check
  their calculations.

- Children then cover the numbers they come up
  with on the 100-grid with counters, aiming
  to cover four numbers in a line.

The mathematics children practise

- Mental addition
- Mental multiplication
- Recall of multiplication facts

Adapting the activity for different ages

Younger children
Help children make a 1–36 number grid or use
the top part of a 100-grid. Do the activity with
two 1–6 dice.

Older children
Use dice with bigger numbers: 1–12, 1–10, 1–20.

Include subtraction and division.

Things to say

*What different numbers could you make with those
dice numbers?*

I wonder if there are any others you could make.

Things to notice

Can the children:

- recall multiplication facts to 5 × 5 instantly?
- recall multiplication facts to 10 × 10 instantly?
- work out multiplications to 5 × 5 mentally?
- work out multiplications to 10 × 10 mentally?

Years 5 and 6

Secret number

An activity about knowing numbers
For 2 to 6 children

You need:

- 100-grid covered in sticky-back plastic (this is useful as children can cross out numbers they have eliminated)
- Felt-tipped pens
- Pencil and paper

The activity

- One child is the Thinker. The Thinker chooses a number from the grid in secret and writes it on a scrap of paper, which is hidden away until the end.
- The other children, the Guessers, take turns to ask questions about the number, to which the Thinker can only respond with 'Yes' or 'No'.

  Is the number more than 50?
  Yes.

  Is it even?
  No.

- The aim of the Guessers is to find out the number by asking as few questions as possible. They can keep track of how many questions they ask and see if they can find the secret number with fewer questions next time.

Variation

Use numbers to 1000, fractions, decimals or negative numbers.

5-minute activities for large groups

When you work with a large group, choose activities that involve everybody so that all the children can join in. Make sure the activities are fast-moving: the children will enjoy the pace, and it will keep them thinking.

The activities in this section are all short and last only five minutes, but you can easily extended or string them together for longer sessions.

The activities are for children in Years 1 and above. If you are working with children in the Reception class, try some of the number rhymes suggested on page 61.

Counting activities

Count forward in ones to 20 (or 10) and back again

What number comes before 6? After 13?

Count round the circle

Count in ones, forward or backward. Children who say a tens number (10, 20, 30, and so on) stand up and drop out of the counting.

Who is going to stand up next?

If we count to 50, how many people will be standing?

How many will still be sitting?

Count in twos from 2 to 30 and back again

What are those numbers called? [even numbers]

Tell me some of the numbers we don't count …

What are they called? [odd numbers]

Count in twos from 1 to 31 and back again

Those are the odd numbers, aren't they?

Do we ever count a tens number this way? Why not?

Count in fives to 30

If we go on, will we say the number 43? How do you know?

Count in tens to 100 and back again to zero

Why is it easy to count in tens?

Count the children in the group

How many eyes/fingers have you got altogether?

How many hands/feet are there in the room? Don't forget your own! And mine!

Games and activities

Pairs to 10

An activity to practise the number pairs that make 10:

10 + 0 **9 + 1** **8 + 2** **7 + 3** **and so on**

- Choose a number such as 4 and say:

  I say 4; you say …

- The group should respond with 6, the number that makes it up to 10.

On the board

Missing number

An activity to practise ordering numbers and mental strategies

- Write the numbers 1 to 10 on the board.
- Ask the children to close their eyes.
- Cover up one of the numbers with your hand.
- Tell the children to open their eyes and ask:

  What number is hidden?

Making 10p

An activity about adding money

- Draw coins on the board: 1p, 2p and 5p.
- Ask the children:

  How can you make 10p, using 1p, 2p and 5p coins?

- Draw children's responses on the board.

  How else can you do it?

  Can you think of another way?

Counting activities

Count in fives from 1 to 51 and back again

51, 46, 41, 36, 31, 26 …

Tell me some of the numbers we don't use.

Count in fives to 99

What would the next number be? How do you know?

Count in tens to 200 and back again to zero

10, 20, 30, 40 …

Games and activities

Pairs to 30

An activity to practise number pairs that make 30

- First, practise some number pairs that make 30: 20 + 10, 19 + 11, and so on.
- Next, choose a number (for example, 14) and say:

  I say 14. You say …

- The group should respond with 16, the number that makes it up to 30.
- Ask one of the children to check the answer with a calculator.

Buzz!

A game to practise facts in the times tables

- Count round the circle in ones. The child who should say 3 says 'Buzz!' instead. As the game goes on, each child who should say a multiple of 3 says 'Buzz!' instead. So the counting goes: *1, 2, Buzz!, 4, 5, Buzz!, 7, 8, Buzz!,* and so on.
- After a while, change the Buzz! numbers to multiples of 5: *1, 2, 3, 4, Buzz!, 6, 7, 8, 9, Buzz!,* and so on.

On the board

Making 12

An activity to practise mental addition

- Write the numbers 1 to 8 on the board.

  How many ways can we make 12 by adding three of these numbers?

- Record children's suggestions on the board.
- Next time, children try making 13, 15 or 20.

Different answers

An activity to practise mental addition and subtraction

- Write the numbers 2 and 5 and the + and − signs on the board.

  How many different calculations can you make, using the numbers 2 and 5 and the + and − signs?

  $$2+5= \qquad 2+5-2= \qquad 2+2+2+2-5=$$

  Call out the answers as I point to the calculations.

Making 50p

An activity about adding money

- Draw coins on the board: 5p, 10p and 20p.
- Ask the children:

  How many different ways can we make 50p, using only the silver coins?

- Record children's suggestions on the board.
- Next time, children try making 60p, 90p or £1.

Counting activities

Count in 20s to 1000 and back again to zero

How many 20s are there in 100? (5)

How many 20s are there in 500? (25)

How many 20s are there in 580? (29)

On the board

Use the board for writing down children's suggestions.

Making 100

An activity to practise mental addition

- Ask children to add any three numbers between 20 and 50 to make 100.

  You cannot use the same number twice, and you must use exactly three numbers.

- Ask one of the children to check.

Different sums

An activity to practise mental multiplication

- Write the numbers 2 and 5 and the $\times$ sign on the board.

  What different calculations can you make, using the numbers 2 and 5 and the $\times$ sign? You can use each number as often as you like.

  $2 \times 5 =$ $\qquad$ $25 \times 2 =$ $\qquad$ $52 \times 5 =$ $\qquad$ $222 \times 2 =$

- Point to the calculations and ask everyone to call out the answers.
- Ask one of the children to check the answer with a calculator.

Guess

An activity to practise mental calculation

- Write a number on the board: for example, 9. Children think up calculations that have that number as their answer. Encourage children to challenge each other if they think a calculation is wrong and get them to check.

Stories, songs and rhymes

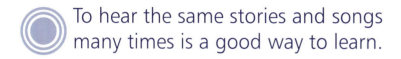

To hear the same stories and songs many times is a good way to learn.

Many stories, songs and rhymes involve mathematics. *Goldilocks and the Three Bears*, for example, involves numbers (chairs, bowls and beds) and sizes (of bears, chairs, bowls and beds). Some songs and rhymes are written specifically to help children learn about counting.

Most of these rhymes are for young children, from the nursery to Year 1 or 2. At this age, children love to hear you repeat stories and rhymes that they know well. You can repeat them in different ways, using number cards and props such as hats to add interest and help with the counting.

Stories

Fairy stories and other tales often have mathematics hidden inside them. When you tell these stories or when children act them out or draw them, you can help them focus on the mathematics by talking about the events and characters.

Goldilocks and the Three Bears

Focusing on height and size

• Ask children to draw the three bears, then cut them out.

How can you make sure that Father Bear is taller than the other two?

Draw Goldilocks, too. She is the same size as Mother Bear, isn't she? Can you make her the same size in the picture?

Putting things in order of size

• Use three teddies to represent the three bears.

Can you say which bear is which? Can you put them in order of size?

Which bowl of porridge is the biggest to give to Father Bear? Which is the smallest for Baby Bear? Which is the medium-sized one for Mother Bear?

Cinderella

- Provide a box full of nearly new shoes from a jumble sale or the charity shop.

Focusing on size

Let's play at Cinderella and see if this shoe fits any of you.

Matching pairs

Here is the shoe that the Prince found. Can you look in this heap of shoes and find its partner?

The Sleeping Beauty

Counting

- Say to the children:

The Sleeping Beauty was asleep for a hundred years. Can you help me count to 100?

Useful tips

When you are reading or telling a story, encourage the children to join in:

- Ask them to show you with their hands the size of Father Bear's bowl.
- Ask them to tell you what the clock looked like when it struck midnight at Cinderella's ball.
- Get them to help you count the number of people pulling at the enormous turnip.

Songs and rhymes

You can use songs and rhymes to teach some mathematics to younger children in a fun and rhythmical way.

Build a House with Five Bricks

Build a house with five bricks,
One, two, three, four, five.
Put a roof on top and a chimney, too,
Where the wind blows through: Whoo-oo-oo whoo-oo-oo …

Children can act this out with their hands: they build the house by placing clenched fists on top of each other; raise arms above the head with fingers touching for the roof; straighten the arms for the chimney – and hoot loudly for the wind.

Once I Caught a Fish Alive

One, two, three, four, five,
Once I caught a fish alive.
Six, seven, eight, nine, ten,
Then I let it go again.

Why did you let it go?
Because it bit my finger so.
Which finger did it bite?
This little finger on the right.

One, Two, Three, Four

One, two, three, four,
Mary at the cottage door,
Five, six, seven, eight,
Eating cherries off a plate.

You might vary this to suit the children you are working with:

One, two, three, four,
Martin at the nursery door,
Five, six, seven, eight,
Eating popcorn off a plate.

Five Currant Buns

Five currant buns in a baker's shop,
Round and fat with sugar on the top.
Along came a girl with a penny one day,
Bought a currant bun and took it right away.

Four currant buns …, and so on

You can act this out with five children pretending to be buns and another child being the person who buys one. Children can also use their fingers to count as they sing.

One, Two, Buckle My Shoe

One, two, buckle my shoe,
Three, four, knock at the door.
Five, six, pick up sticks,
Seven, eight, lay them straight.
Nine, ten, big fat hen,
Eleven, twelve, dig and delve.
Thirteen, fourteen, maids a-courting,
Fifteen, sixteen, maids in the kitchen.
Seventeen, eighteen, maids in waiting,
Nineteen, twenty, my plate's empty.

Six Little Snails

Six little snails
Lived in a tree.
The wind blew hard
And down came three.
How many were left?
Three little snails
Lived in a tree.
The wind blew hard
And down came three.
How many were left?

The children can show numbers with their fingers. You can make up variations. For example: *Ten little snails lived in a shoe, the wind blew hard and out came two.*

Useful tips

Children can use their fingers to show which number they are up to.

- With subtraction songs such as 'Six Little Snails', children show the number with their fingers. They can bend down the number of fingers they need to subtract. Ask: *How many are left?*
- Instead of saying the numbers, point to them on a number line and let the children say them for you.
- Use props such as pictures of currant buns or toy currant buns to make the rhyme real.
- Children can hold up number cards to show the numbers.

Cooking

Following a recipe

With younger children, you need to help them read, interpret and follow the recipe. The task for older children is to read the recipe themselves and do the cooking on their own. You can check from time to time by asking them questions about what they have to do next. An extra challenge for older children is to ask them to increase the quantities in the recipe to make enough for two groups or the whole class.

The mathematics in cooking

Weight   **Using a balance with weights one side or kitchen scales**

It is important to let children do as much weighing and measuring as possible themselves. Young children will weigh things using a balance. This can be difficult because the arms of the balance have to be level. Often, young children do not have the experience to know how to do this. They keep piling things on one side when they should be taking them off so that the balance stays firmly down on one side. It takes patience to watch them and let them do it before they finally realise what they have to do.

Young children will also spill ingredients. They will pile flour onto the balance and then see that they have to take lots off again. They will over- and underestimate quantities in surprising ways. But it is vital that they do all this themselves so that they can learn from experience.

When older children use kitchen scales, they sometimes need help in reading the dials. You can point out the range of numbers and ask children to work out what numbers the markers in between stand for.

Capacity   **Using a measuring jug or measuring spoons**

You also need to help children with measuring jugs and spoons. If children have not yet done much work with measuring jugs, it may be best just to tell them where the right mark is. You could explain what the numbers mean. Encourage older children to read and make sense of the numbers on jugs and to work out what the markers in between two numbers mean: for example, the line halfway between 100 ml and 200 ml is 150 ml, but it may not be numbered.

Time   **Using a clock or a kitchen timer**

A kitchen timer is useful to show children how to measure the passing of time. The timer shows you at any point how much longer you have to wait. You can use this alongside a classroom clock and compare the two. For example, you could say: *How many more minutes will the buns take to cook? Seven? Look at the clock – when will seven minutes be up? What time will that be?*

Number

You can ask young children questions about how many rolls or cakes they are making, how many they have made so far and how many more they need to make. Encourage the children to arrange the buns or biscuits in rows so that they can see the pattern of the numbers: 5, 10, 15, 20 …

They can count in twos or threes or show you half the buns on the tray.

Soda bread buns

Ingredients for 12 buns

150 g plain white flour
150 g wholemeal flour
2 teaspoons baking powder
1 heaped teaspoon sugar
A large pinch of salt
40 g butter
A squeeze of lemon juice
200-300 ml milk
Flour for dusting

To serve:
Butter, sliced banana

The method

- Preheat the oven to 190 °C (gas mark 5).
- Sift the dry ingredients together into a large bowl.
- Rub in the butter.
- Add the lemon juice and mix in enough milk to make a soft dough. The dough should not be too wet.
- Divide the dough into 12 portions and shape each portion into a bun – don't handle the dough too much. Leave the buns with a rough finish.
- Place the buns on a greased baking sheet.
- Dust them with a little flour and make a small cross in the top.
- Bake them for 10-15 minutes or until the buns are golden brown.

Things to say

Whatever recipe you use, the questions to ask children and the things to notice about what they do will be similar.

Which of the weights do you need to weigh out 150 g of flour?

Where on the scale is the 150 g mark? How can you tell?

Can you show me where on this jug is 200 ml? How do you know that?

What does 'heaped teaspoon' mean?

Do you think all the rolls are exactly the same size?

How many buns have you made? How many more do you need to make 12?

At what time will the buns be ready to take out of the oven?

How long should we set the timer for?

Things to notice

Can the children:

- follow a recipe?
- weigh out the ingredients accurately, using a balance of some sort?
- read the dial of the kitchen scales accurately?
- read a capacity container accurately?
- work out the passing of time?
- work out the approximate amount of ingredients they need for 24 or 36 buns (or six buns)?

The home corner

The home corner is an area that you can adapt for all kinds of role play. Sometimes, the home corner is a pretend home, and at other times, you can turn it into a hospital, a shop or an aeroplane … The home corner is useful in helping children with their social and language skills, and there is also plenty of scope for mathematics. As young children get completely absorbed in their play, make sure that the maths learning fits into their imaginary games.

Occasionally, you can join the children's games by taking part in their fantasy. You can then ask questions to make children think, such as: "The teddies and I would all like a cup of tea, too. Are there enough cups?" or "Panda wants to go to sleep as well. Is there a bed big enough for him?" But don't be surprised if children give an answer that seems 'wrong': they may not mind if the bed is too small for Panda or if there are more saucers than cups – you just have to acknowledge that their imagination does not focus on anything mathematical at that particular moment.

It can work well to set up an activity away from the home corner that is linked to it, and which has a mathematical theme. Often, children will take these ideas back to the home corner in their play. For example, you could put out dolls and dolls' clothes on a table. The dolls and the clothes are all different sizes, and the children have to find the right clothes for each doll. Or you could set up a telephone kiosk: children could invent numbers and make a telephone book. When you put these activities back into the home corner, children will often merge them into their own games.

Useful tips

- Young children can get very involved in their role play in the home corner, and it does not always help to interrupt this by talking with them about what they are doing or suggesting a new activity. The children's enjoyment and interest must come first, so your ideas will sometimes have to give way to what the children want to do.
- Try observing for a while, then enter the game by taking on a role yourself.
- Set up an activity while the children are still on the mat with the teacher and invite some of them to join in.

Train, shop, hospital, café ...

You can turn the home corner into a place of work such as a train, shop, hospital, café, and so on. This gives scope for counting and using money as well as reading and writing numbers.

Related activities to set up away from the home corner

Children can:

- talk about the time for their appointment and draw a clock to show that time.

- make travel tickets with you, showing the price and the seat number.

- count out coins to pay the shopkeeper or ticket collector.

- write menus and price lists.

- make price labels to stick on the goods for sale.

- weigh their luggage on bathroom scales.

- take each other's temperature and write it down.

- put out the seats for the train and number each one.

Things to say

How many people can go on the train?

How much does it cost to go to London?

How much does your suitcase weigh?

What number is your seat on the train?

How many people are waiting to see the doctor?

What time is your appointment?

How high is her temperature?

Having a picnic

There is lots of mathematics in real or imaginary meals, parties and picnics.

Related activities to set up away from the home corner

Children can:

- lay places for all the teddies.
- put a 'teabag' in each mug.
- put a cherry on each cake.
- share out the biscuits fairly.
- choose a bottle that they think holds enough juice for four people and then check whether it does.
- compare two jugs to see which holds more.

Things to say

How many buns are there? I wonder if there are enough for everybody.

Which jug do you think holds more? How can we check?

You cut the sandwiches exactly in half. How many pieces are there now? What if these pieces were cut in half again?

Can you give the biggest bun to the biggest teddy?

Washing up

When the children do pretend washing up or wash real plates or equipment with you, they get experience in counting and in comparing the sizes of things.

Things to say

How many mugs can we fit in the bowl? Let's count: 1, 2, 3, …

I wonder if the forks will fit in this box.

Can you count how many spoons there are here?

Count together with Alisha how many dirty plates there are in that pile.

Let's sort the things we have washed up and put them away.

Will you help me arrange the clean pots in order of size?

Out and about

On an outing

When you are out and about with children – in the playground or on an outing – you can use mathematical language and ideas. Children are naturally interested in the everyday world, so you can supply them with facts and ideas to think about.

How many paving stones can you jump?

What time does the clock say?

Which is taller, the bus stop or the house?

How much are the bananas at the market?

Useful tips

- Help children learn how to tell the time. One child can keep an eye on the time (either their own watch or a school clock) and tell you when playtime is over.
- Each day, choose a different theme to interest the children: for example, encourage them to count hops, skips and jumps during playtime; work out how many wheels there are on the staff cars; or guess, then check, how many strides they need to get to the far wall and back to you.

Just talking

- Discuss the route to the shop, the school garden or the main entrance.

  Is the road straight? Or curved?

  Do we turn left at the corner or do we go straight on?

Counting

- Count children, swings, tyres, dogs, lamp posts or trees. Discuss whether there are more trees or lamp posts.
- Count the tricycles.

  How many wheels on a tricycle? On two tricycles? On three?

- Count the pushchairs.

  How many wheels on a pushchair? On two pushchairs? On three?

- Look at the windows in the houses you pass.

  What different shapes can you see?

  Do all the windows have the same number of panes?

  How many window panes are there in that window? And in that one?

Numbers

- Look for numbers on doors.

  What do you notice about the door numbers in a street? Why do the numbers go 1, 3, 5, 7 …?

- Look for numbers on buses, in shop windows, on posters or street signs.

  Who can find a sign with their age in it?

- Read out the numbers on car number plates.

  Who can find a number plate with the number 3 in it?

  Which number in that number plate is the biggest?

Time

- Look out for clocks on shops and other buildings. See if the children can tell the time; if not, read them the time yourself.

  What time does the clock say? Is it right?

- Discuss what sort of display the clock has: analogue (with hands) or digital (with numbers only)
- Talk about what sort of numbers the clock has: numerals made up of light-bars Arabic numerals (1, 2, 3, and so on) or Roman numerals (I, II, III, IV, V, and so on)

  Do you know what those Roman numbers mean?

Patterns

- Look for patterns in brickwork, fences or paving stones. Look at the patterns in the way tins are arranged on a shelf. Notice fabric on sale in shops.

  How would you make that pattern yourself?

  Can you see where the pattern repeats?

  Why don't these tins fall down?

  How is that pattern different from that one?

Post boxes

- Find out who can reach the letter slot.
- Discuss which is taller, the post box or the wall.
- Look to see how many times the post van collects letters each day.
- See if children can say the collection times in two ways: for example '3:45' and 'a quarter to 4'.

  Is the post collected at the same time every day?

  At what times are the collections made?

Bus stops

- Explore bus stop signs and timetables.

  Which buses go from this stop?

  What does the timetable tell us? When is the next bus?

Opening times

- Shops, doctor's surgeries, swimming pools, and so on, often display their opening times on a notice outside their door. Look to see what time they open and close.

  How many hours is the pool open?

  Does the museum close at all during the day? For how long?

Position

- Talk about where things are to encourage children to use the language of ordering, of direction and of position.

  What is next to the sweet shop?

  What is between the post office and the greengrocer's?

  Can you remember which is the first shop we come to? And the next one?

  What is under/behind/in front of the tree?

Collections

- Collect conkers, acorns and fir cones to use as counting materials back at school.
- Collect feathers and conkers for weighing.

  How many conkers have you got there?

  Have we got more conkers or fir cones?

  These feathers are so light I can hardly feel them.

- Keep bus and train tickets.

  What numbers can you see on the ticket? Do you know what they mean?

  Can you see the date there? What does 21/11/09 mean?

In the playground

In this section, you will find games and activities involving mathematics which you can help children organise in the playground.

About the games and activities

The games and activities are for three different age groups. However, you could try the games with children of other ages and see what appeals to them. Sometimes, you may be supervising a lot of children outside. In this case, teach a group of children a game and then leave them to play by themselves. At times, you can work with another adult and take turns playing with a small group of children and taking responsibility for the rest of the group.

Counting hops, skips and jumps

Children need to practise counting. Activities they enjoy give them a purpose for counting.

Children might use the following:

skipping ropes	beanbags
balls	skittles
sand timers	hoops

Things to notice

Can the children count:

- how many times they can hop, jump or skip without stopping?
- how many hops, jumps or skips they can do before the sand in the timer has run through?
- how many times in a minute they can throw a beanbag at a skittle or into a hoop?
- how many times they can throw and catch a ball without dropping it, sseither on their own or in pairs?

Things to say

What number comes next?

How many skips did you do? I wonder how many you will be able to do next time.

See how many hops you can do along this line.

What number do you get to if you skip across the playground, counting your skips?

Can you do 100 throws? It doesn't matter if you stop: you can carry on counting without starting at 1 again.

Who did the most hops?

I wonder on which of your legs you can hop for the longest.

How many skips do you think you could do in half a minute?

How many more hops do you need to do to reach 20?

How many more jumps did you do this time?

Using playground markings

Many playgrounds are painted with grids, ladders or other shapes. Sometimes, numbers are painted on the playground. You can use these markings for all sorts of mathematics. Sometimes, you may need to improvise.

Number track (straight, wavy or spiral)

Children can:

- walk along the track, saying each number as they step onto it.
- find a number with a 6 in it, then find another one.
- start on 1 and jump on every other square, saying the numbers they land on.
- start on 2 and jump on every other square, saying the numbers they land on.
- work out how many jumps of two they need to make to get to 10.
- stand on a number to hide it and challenge a friend to say what the hidden number is.
- run along the lines clockwise, then stop and go anticlockwise.

Things to say

Walk along the track, saying the numbers … Now stop and close your eyes. Can you tell me the next few numbers?

Stand beside me and close your eyes. Can you point to where the 10 is?

How many numbers are there with a 5 in them? With a 1 in them? Why are there more 1s than 5s?

100-grid

Children can:

- use a giant dice and stones instead of counters to play a 'track' game where they race each other from 1 to 100.
- count in threes and put a skittle on every third number. Discuss the pattern the skittles make. (They will make diagonal lines on the grid.)
- put a beanbag on all the numbers in the 5 or 10 × table. Get friends to stand on all the numbers in the 4 × table. Discuss which columns don't have any beanbags (or friends) on them, and why.
- throw two beanbags onto the grid, then predict how many numbers they must count on from one beanbag to the next beanbag or how many numbers there are between the two beanbags.

Things to say

You are on 47. Where will a dice roll of 4 take you to?

How many steps to get to 80 from where you are?

Is the number you are on in any of the tables you know?

Go and stand on any number in the 3 × table.

Making shapes

Children can:

- work with a friend to make a shape with their bodies. Other children get into pairs and copy the shape.
- work with a friend to make a number or letter shape for you to guess.
- pretend they are a mirror, then face a partner and copy their movements.

Things to say

How many straight lines do you need to make an E?

Could the four of you make a whole word with your bodies?

Giant strides

Draw a line for children to stand on and a second line some distance away across the playground (or use the playground markings if they are already there).

Children can:

- say how far they think it is from one line to the other in giant strides, then test out their estimates.
- estimate the distance in metres, then check with a surveyor's tape or metre rule.
- predict and then find out who takes the fewest strides.
- work out how many strides it will take to get across and back again.

Things to say

How many more strides did you take than Nasrul?

Mina took 23 strides. Do you still think your guess of 50 strides is a good one or would you like to change it?

How many strides will it take if you go there and Alan comes back?

Is your giant stride longer than a metre? Is it shorter than a metre?

Timing

Children can:

- count in tens while you time them; see how high they can count in a minute.
- time a friend as they run right round the playground.
- check how many times their friend's heart beats in a minute, before and after exercise.

Things to say

You ran round faster than Jasmin, but did you run as far? Was it fair?

How will you know when a minute has passed?

Estimating

Children can:

- make a sensible guess as to how many bricks there are in the wall.
- estimate how many children there are in the playground.
- say how many metres high or tall they think the tree/climbing frame/school is.

Things to say

How might you work that out?

How many bricks are there in a row? And how many rows are there?

How many children are there just in this small area? Does that help you work out how many there might be in the whole playground?

Show me a metre with your hands.

4 Useful information

Maths equipment

There is usually a lot of equipment in the classroom to help children understand mathematics. It is not always obvious what each piece of equipment is for. Even when it is easy to see what it is for, you may not know how to use it to help children learn. In this section, we talk about the equipment you are most likely to come across and how to use it.

Number

Base 10 blocks

Often called Dienes or Multibase blocks. The blocks are either wooden or plastic and are used to help children understand place value (see Glossary, page 88) or what used to be called 'hundreds, tens and units'. They have small cubes known as 'units' or 'ones', 'longs', which are 10 units long, 'flats', which are 10 longs wide, and 'blocks', which are 10 flats high.

Calculators

Children need to know quite a lot of mathematics to use a calculator properly. They have to decide what kind of calculation they are doing and which buttons to press. If they press the wrong button and get a wrong answer, they should be able to spot this and start again. The calculator should not be used to avoid learning number facts. Children need to know their number facts and be able to calculate in their heads.

Cards

Number cards from 0 to 9 and from 0 to 100 are popular and useful for many games and activities. You can also use ordinary playing cards for many number games and activities.

Coloured rods

Cuisenaire and Colour Factor are the most common types of coloured rods. These are wooden rods of different lengths, representing numbers 1 to 10 or 12. Use the rods for addition, subtraction, multiplication and division as well as fractions.

Counters

Counters, conkers, fir cones, beads, straws and buttons are all good for counting. Children like unusual or attractive materials, and these make counting activities thoroughly enjoyable.

Cubes

You can join Unifix cubes together to make sticks of various lengths. These cubes are useful for counting, place-value work (see Glossary, page 87) and block graphs.

You can join Multilink together on all sides. Multilink cubes are useful in the same way as Unifix, and you can also use them for work on volume and shape.

Centicubes are 1 cm wide, long and high. You can join them together in any direction. They are useful for work on capacity and number. Each centicube weighs 1 gram.

Dice

You can find 1–6 dice in most classrooms. Use them for games and mental mathematics activities. The school may also have different number dice with four, eight, 12 and 20 faces.

Blank dice are useful because you can write numbers of your choice on them.

Dominoes

Use dominoes with younger children for counting and matching, and with older children for all kinds of number puzzles. You can get dominoes up to double nine, as well as the more usual double six.

100-grid

These 10 by 10 grids have the numbers in order from 1 to 100. They are useful for all sorts of activities such as counting in tens from any number (5, 15, 25, 35 …) or adding 9 by adding a 10 and going back one space.

Multiplication grid

These 10 by 10 grids contain all the multiplication facts from 1×1 to 10×10. They help children develop rapid recall of times table facts to 10×10. Children can use multiplication grids individually or in paired work in maths problems and activities.

Number lines

Number lines help children build up an image of numbers in order and help them learn to count forward and backward. Use number lines for multiplication, division and number patterns. There are also get blank number lines, to help children with larger numbers, fractions or decimals.

Numicon

Numicon are a set of coloured plates and pegs that provide children with a visual and tactile image of numbers and number relationships: number order, odd and even, combining and splitting numbers, number pairs that make 10, and so on.

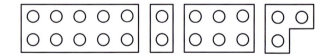

Capacity

Measuring jugs
You can get measuring jugs of different capacities such as one litre, two litres or $\frac{1}{2}$ litre.

Jugs and cylinders are calibrated in different ways. Children need to learn how to read the scales, which some find tricky.

Money

Coins and notes
In some classrooms, children use toy coins and notes, while in others, they use real coins. Using coins in maths activities gives children practice in number calculations in a context they are familiar with from their everyday lives.

Length

Rulers and metre rules
Rulers and metre rules are marked in different ways. They may be marked in blocks of 10 centimetres, in centimetres and half centimetres or in centimetres and millimetres. Children need to learn that they must start with the 0 point when measuring length, which is not necessarily at the beginning of the ruler.

Surveyor's tape
These are usually 10 m or 20 m long. They are marked in metres and decimals of a metre. The markings can be difficult to understand, so you may find them more suitable for use with older children.

Tape measures
Tape measures are useful for measuring curved objects or things measuring more than a metre. If one side is marked in feet and inches and the other in centimetres, make sure that children are using the correct side (centimetres).

Measuring angles

Protractors, angle indicators and rotograms
These are all used for measuring angles. An angle is an amount of turn and is measured in degrees: for example, a full turn is 360°, and a quarter turn or right angle is 90°.

Weight

Balances
Balances compare the weights of two objects. You can also use weights on one side to balance an object.

Weighing scales
Like old-fashioned kitchen scales, weighing scales have a pan on one side and a platform for weights on the other. They work in the same way as a balance. Use metric weights (grams and kilograms), not pounds and ounces.

Dial scales (bathroom and kitchen scales)
Children can find dials difficult to read. Only use dial scales when children have had experience of balances. Again, use metric units (grams and kilograms), not pounds and ounces.

Spring balances
The more something weighs, the more the spring is pulled down. Some spring balances are used to measure forces in science and are calibrated in Newtons. Others show grams and kilograms. Again, you will need to explain to children how to read the scales.

Time

Digital clocks

Children need to learn how to read digital clocks as well as the traditional (analogue) type of clock.

Geared clocks

A geared clock is more realistic than a simple card clock face, as the hands move together like a real clock. If you prefer, use an old household clock from a charity shop or jumble sale.

Sand timers

These are just like old-fashioned egg timers. Use them to measure specific lengths of time such as three minutes.

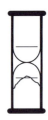

Kitchen timers

Use a kitchen timer to measure a specific length of time. The timer will ring when the set time is up. This is useful for giving children a sense of how long a minute, five minutes or half an hour is.

Stopwatches

These measure time extremely accurately. Use them with older children to compare running speeds.

Temperature

Thermometers

Make sure the children are using thermometers that are calibrated in Celsius (also known as centigrade), rather than Fahrenheit. You can use thermometers for science as well as mathematics.

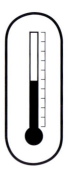

2D shapes and 3D solids

Flat shapes

These are used for activities such as sorting and pattern making. They are useful for getting to know the names of 2D shapes and what they are like: for example, squares fit together with no gaps and circles don't.

Solid shapes

3D solids are used in activities such as sorting. Again, these are useful for getting to know names and properties of solids: for example, cubes have six sides (faces), all of which are square.

Glossary

Analogue
Analogue clocks and watches show the time using hour and minute (and sometimes second) hands.

Bar chart
A bar chart is often used in primary schools to display information about numbers of people or things: for example, the number of children who voted for different names for the new guinea pig.

Capacity
The capacity of a container is the amount of space inside it.

Centimetre
A centimetre is shorter than half an inch. There are 100 centimetres in a metre.

Data, data handling
Data is another word for 'information'. When children work at data handling, they collect information, then organise and display it in some way. Children might collect data about people's heart rates or running speeds.

Difference
The difference between two numbers is how much bigger or smaller one is than the other. If Sammi has five marbles and Harun has nine, the difference is four – in other words, Harun has four more than Sammi, or Sammi has four fewer than Harun. You can find the difference by subtracting one number from the other.

Digit
All numbers are made up using the digits 0, 1, 2, 3, 4, 5, 6, 7, 8, 9. Two-digit numbers such as 56 have two digits; three-digit numbers such as 398 have three; one-digit numbers such as 7 or 3 have one.

Digital
Digital watches and clocks show the time in the form of digits.

Even
The even numbers (2, 4, 6, 8, and so on) can all be divided exactly in two.

Factor
When a number is broken up into smaller numbers by division, these smaller numbers are called factors. For example, 30 can be broken up into two of its factors, 5 and 6, because $5 \times 6 = 30$. 3 and 10 are also factors of 30, because $3 \times 10 = 30$. So are 15 and 2, because $15 \times 2 = 30$. The final two factors of 30 are 1 and 30 as $1 \times 30 = 30$.

Metre
A metre is just over a yard or 3 feet. There are 100 centimetres in a metre.

Model
A model of something shows its essential character, but is not an exact replica. A scale model of a house shows the shape exactly as it is, but much smaller, and may not bother about details such as showing each brick. You can use a number line to model numbers in much the same way. A number line shows some of the essential facts about the number – the order of numbers, and how far each one is from 0 – but does not tell you whether the number is odd or even or whether it can be divided by 5.

Multiple
If you multiply 9 by other numbers, the results are all multiples of 9. So, for example, 18, 27, 90 and 99 are all multiples of 9. Similarly, 6, 24, 60 and 660 are all multiples of 6.

Multiplication tables
Some children learn their multiplication tables by chanting. Other children learn the multiplication facts a few at a time, perhaps by looking at the number patterns in the tables. How they learn does not matter as long as they do learn the multiplication facts.

Negative numbers
Numbers less than zero such as -3 or -10 are called negative numbers.

Net
When you unfold a box and lay it flat, you are making the net of the box. Children often explore different nets that can all be folded up to make a cube.

Number bonds

Number bonds or number facts are the basic facts about number which children need to recall rapidly. Important addition and subtraction bonds include $2 + 2 = 4$, $2 + 3 = 5$, $8 - 6 = 2$, $8 - 7 = 1$. The important multiplication and division facts are all in the multiplication tables.

Number line

A number line is a useful way of modelling numbers. Number lines start from 0, not 1. Children can work out addition and subtraction problems by drawing steps and jumps from one number to another. Older children can look at the spaces between the numbers. They can talk about the decimal or fraction numbers that belong in those spaces.

Number tracks

Number tracks are useful for young children and appear in games such as Snakes and Ladders. Children can move counters along them and begin to learn how many steps it takes to get from one number to another. Unlike number lines, number tracks usually start from 1.

Odd

Odd numbers (1, 3, 5, 7, and so on) are numbers which you cannot divide exactly by two.

Operations

$+ \quad - \quad \times \quad \div$

There are four main number operations: addition, subtraction, multiplication and division. When you start with a number such as 59 and add or subtract something from it, you are operating on the number 59.

Place value

In any number over 9, the digits have different values depending on their position, or place, in the number. So, for example, in the number 749, the 7 is worth 700, but in 73, the 7 is worth 70. This is called place value.

Product

When you multiply two numbers together, the result is called the product. The product of 3 and 4 is 12.

Quotient

The quotient is the answer to a division problem. For example, when you divide 20 by 5, the quotient is 4.

Remainder

What is 'left over' when doing a division problem. For example, when you divide 21 by 5, the answer is 4 remainder 1. This can be written as: $21 \div 5 = 4$ r 1

Subtract

Subtraction is a general word that covers both 'take away' and 'difference'. If Janice has nine marbles and loses five, a subtraction or take-away calculation will tell us that Janice has four left. If Jamie has nine marbles and Janice has five, the difference is four: Jamie has four more than Janice. These two situations are quite different. In the first situation, only one person has marbles. In the second situation, both people have marbles. But both these situations can be written out as the same subtraction calculation: $9 - 5 = 4$

Sum

In ordinary speech, a 'sum' can be an addition, a subtraction, a multiplication or a division. In maths, the term 'sum' should strictly be used for additions only, because it actually means 'total'. So the sum of 5 and 9 is 14. The term 'calculation' can be used to refer to any of the four operations.

Unit

Unit means 'one'. It describes whole numbers under 10 as in the phrase 'hundreds, tens and units'. Units are also used in measuring. When we measure weight or length, we use units which are all the same size, such as inches, centimetres or grams. You can put these units together to make larger units (for example, 12 inches to the foot or 1000 grams to the kilogram).

What most children should be able to do
at the end of Years 1, 2, 3, 4, 5 and 6 …

Maths targets

This section shows some of the key things that children of different ages should know at the end of each year. It is likely that you will work with some children who cannot do everything on the list for their age group. But it is important to know what we expect of children and to help them achieve it.

By the end of **Year 1** most children should be able to …

- count 20 or more objects
- read and write numbers to at least 20
- put numbers to 20 in order
- say the number one more or one less than another number
- say the number 10 more or 10 less than a number in the 10 × table
- count in ones, twos, fives and tens
- add and subtract numbers to 5 in their heads
- recall rapidly pairs of numbers that make 10 (for example, 6 + 4)
- work out problems like 'How many wheels are there on three bikes?'
- work out doubles of numbers to double 5
- describe the position of objects
- solve simple problems about money (for example, work out how to make 14p with three coins)
- read the time to the hour or half hour on a clock face.

By the end of **Year 2** most children should be able to …

- count and read numbers to at least 100
- count to 100 objects by putting them in groups of 2, 5 or 10
- put numbers to 100 in order
- round two-digit numbers to the nearest 10
- add and subtract numbers to 10 in their heads
- work out or recall rapidly pairs of 'tens numbers' that make 100 (for example, 60 + 40)
- work out doubles of numbers to double 10
- find half, a quarter or three quarters of a set of objects
- work out or remember multiplications and divisions in the 2 ×, 5 × and 10 × tables
- recognise reflective symmetry in patterns and shapes
- solve problems with money (for example, find ways of spending 50p using silver coins)
- read the time to the quarter hour.

By the end of **Year 3** most children should be able to …

- read, write and order numbers to 1000
- round three-digit numbers to the nearest 10 or 100
- read and write fractions such as $\frac{3}{7}$ or $\frac{9}{10}$
- add and subtract numbers to 20 in their heads
- work out or recall rapidly some pairs of numbers with a total of 100 (for example, $35 + 65$)
- work out or remember multiplications and divisions in the $2\times$, $3\times$, $4\times$, $5\times$, $6\times$ and $10\times$ tables
- multiply numbers by 10 or 100 in their heads or on paper
- use the four compass directions (N, S, E and W)
- solve problems with money, such as finding five coins to make 74p
- read the time to the nearest five minutes.

By the end of **Year 4** most children should be able to …

- read, write and order numbers to 1000
- begin using negative numbers
- read and write decimal numbers such as 3.5 and 9.02
- recognise fraction and decimal equivalents (for example, $\frac{1}{2}$ is the same as 0.5, and $\frac{1}{10}$ is the same as 0.1)
- work with 'mixed numbers' such as $3\frac{1}{2}$ and know where they belong on a number line
- add and subtract numbers to 100 in their heads
- multiply or divide whole numbers by 10 or 100 in their heads or on paper
- work out or remember multiplications and divisions in any table to 10×10
- work out, for example, $41 \div 4$ as 10 remainder 1
- find fractions of numbers of objects (for example, find $\frac{1}{5}$ of 30 plums)
- recognise horizontal and vertical lines
- solve problems with money, such as buying three different things and working out change for £10
- read the time to the nearest minute
- use a ruler to draw and measure lines to the nearest centimetre.

By the end of **Year 5** most children should be able to …

- count in ones or tenths, backward and forward
- read, write and order whole and decimal numbers
- recognise fraction and decimal equivalents (for example, $\frac{7}{10}$ is the same as $\frac{14}{20}$ and 0.7)
- understand and use simple percentages
- multiply or divide whole numbers and decimals by 10, 100 and 1000
- continue to know multiplications and divisions in any table to 10×10
- mentally add and subtract numbers such as $6070 - 4097$ or $4.5 + 3.2$
- find pairs of factors of two-digit numbers (for example, 5 and 6 are factors of 30 because $5 \times 6 = 30$)
- find fractions using division (for example, find $\frac{1}{100}$ of 5 kg)
- solve problems with money, such as finding the total cost of several items, then working out the change from £20
- measure angles with an angle measurer or protractor
- use a timetable to work out how long a journey lasts
- work out the perimeter and area of a rectangle.

By the end of **Year 6** most children should be able to …

- find the difference between a negative and a positive number (for example, the difference between -3 and $+17$)
- read, write and order whole numbers and decimal numbers such as 3.409
- simplify fractions such as $\frac{3}{60}$ by cancelling common factors
- order a set of fractions
- identify prime numbers to 100
- find the squares of numbers to 12×12 and the squares of numbers such as 20 or 50
- multiply and divide with decimals (for example, 0.8×7 or $4.8 \div 6$)
- work out, for example, $63 \div 8$ as 7.875
- solve problems with money, such as sharing £363 630 among three people or the sale price of some £30 jeans with 10% off
- calculate angles in a triangle
- convert between units of measurement (for example, change 2.75 litres to 2750 ml)
- use a ruler to draw lines and measure to the nearest millimetre.

Feedback sheet

A feedback sheet is a useful way for the adults in a busy classroom to communicate.

Teaching assistants are often asked to support a child or a group of children with a maths task set by the teacher. A feedback sheet is a useful way for the adults in a busy classroom to communicate, as finding a chance to talk can be difficult. You can change the sample feedback sheet on page 95 to suit your school's own needs.

How to use the feedback sheet

Before the lesson

The teacher fills in:

Activity
What the activity involves

Equipment needed
What equipment will be needed

Words to use
Any specific mathematical vocabulary that the teaching assistant should use and encourage the children to use

Key learning points
Up to three things that the teacher particularly wants the teaching assistant to look out for (for example, 'Does the child add numbers mentally? Does the child read the weighing scales dial correctly?')

The teaching assistant:

- reads what the teacher has planned
- checks that the equipment is available.

During the lesson

The teaching assistant:

- discusses with the children what they are doing
- asks open questions to prompt children's thinking and encourages them to demonstrate what they know and understand
- asks closed questions to find out what facts children know
- ticks the 'Can do' or 'Needs help' boxes to show how children did at each key learning point
- comments on anything else that would interest the teacher.

After the lesson

The teaching assistant:

- discusses with the children what they have learnt.

For the teacher

Feedback sheet

Name _____ Date _____

Activity

Equipment needed

Words to use

Key learning points

1 _____

2 _____

3 _____

For the teaching assistant

Children

	Can do	Needs help	Comments
1 **2** **3**	☐ ☐ ☐	☐ ☐ ☐	
1 **2** **3**	☐ ☐ ☐	☐ ☐ ☐	
1 **2** **3**	☐ ☐ ☐	☐ ☐ ☐	
1 **2** **3**	☐ ☐ ☐	☐ ☐ ☐	
1 **2** **3**	☐ ☐ ☐	☐ ☐ ☐	
1 **2** **3**	☐ ☐ ☐	☐ ☐ ☐	

Other resources

Jills of All Trades: Classroom Assistants in KS1 classes

Janet Moyles and Wendy Suschitzky

ISBN 978 0 9029 8389 2

This book is the result of a 16-month investigation into the roles and relationships of KS1 teachers and the assistants they work with, including those training as Specialist Teacher Assistants. It outlines existing practice in primary schools and makes recommendations for future development. A summary of the research and recommendations is also available.

Help in the Classroom

Margaret H Balshaw

David Fulton Publishers

ISBN 978 1 8534 6476 8

This book provides practical advice to help your school make the most of its teaching assistants. It looks at the issues surrounding the employment and training of non-teaching staff and suggests ways to examine and develop school policies. The book is aimed at coordinators and those responsible for in-service training, rather than teaching assistants themselves.

Teaching Assistant's Handbook

Teena Kamen

Hodder-Arnold

ISBN 978 0 3409 5938 1

A practical handbook aimed at supporting TAs through the various NVQ and STA courses and qualifications. It covers all the key aspects of classroom, behaviour and resource management, and has good sections on learning and the acquisition of skills. It does not focus mainly on maths.

A Toolkit for the Effective Teaching Assistant

Richard Tyrer, Stuart Gunn, Chris Lee, Maureen Parker, Mary Pittman and Mark Townsend

Paul Chapman Educational Publishing

ISBN 978 1 4129 0061 4

This book is designed to help teaching assistants to be more effective in their practice and to develop their professional capabilities. The chapters are designed to allow readers to dip into and out of the book as necessary. Areas covered include: roles and responsibilities of teaching assistants; working as a team; understanding pupils' learning styles; self-esteem and how it links to classroom behaviour.

Mathematics Accomplished: the BEAM Booster Packs

BEAM Education

Y1: Code 6416931
Y2: Code 6416932
Y3: Code 6416933
Y4: Code 6416934
Y5: Code 6416935
Y6: Code 6416936

These pairs of books, for each year from Y1 to Y6, help teachers and teaching assistants work together to support children who need additional practice in maths. Each teacher's book has 30 lessons; the accompanying teaching assistant's book has a further activity for each lesson, for those children who need a bit more help and practice.

The Know-All Boxes

Fran Mosley

BEAM Education

The Know-all Number Bonds Box
ISBN 978 1 9062 2464 6

The Know-all Times Tables Box
ISBN 978 1 9062 2465 3

There are two packs, one on addition and subtraction number bonds and one on times tables, for teaching assistants to use with small groups of children who have not yet learnt the necessary number facts. Each pack contains all the material needed for running 40 sessions: plenty of games, all the necessary equipment and a CD-ROM with tests and achievement certificates.

Essential Maths Workout

Jenny Murray

BEAM Education

ISBN 978 1 9031 4294 3

This is a collection of 24 colourful and enjoyable games for teaching assistants to use with 7- to 9-year-old children who need extra practice in key aspects of maths (not just number). All the equipment needed is included, and there are notes and recording sheets to help the teaching assistant feed back to the teacher.